Foreword

The origin of this book is a report entitled 'Wear Resistant Surfaces—a guide to their production, properties and selection', prepared by IRD staff in 1976. Those mainly involved were Dr G Arthur, Mr D Birch, Mr G M Michie, Mr P Moorhouse and Dr T C Wells who was the project leader. Dr T Bell, then at Liverpool University, acted as a consultant for the chapter on thermochemical treatments. The work was financed by about eighty sponsors, approximately half from the United Kingdom and half from North America, Australia and Europe.

In view of the increasing number of types of surface treatment available for reducing wear and their increasing importance in the engineering industry, IRD proposed to the Process Plant Committee of the Department of Trade and Industry that the 1976 work should be brought up to date, extended, and made widely available. This proposal was accepted and financial support was provided by the Department. The revision was carried out during 1984 by Dr G Arthur (project leader and editor), Mr D J Brown, Mr P Moorhouse of IRD; Dr J Edwards, a private consultant, and Professor T Bell of Birmingham University.

International Research & Development Co Ltd
April 1985

Acknowledgements

Tables and illustrations are reproduced by the kind permission of the following authorities.

AIME: Figures 5.9, 9.11 and 9.12; Table 9.9.

American Institute of Physics: Table 9.3.

American Machinist: Table 3.14; Figure 3.22.

American Nuclear Society: Copyright 19— (date as appropriate) by the American Nuclear Society, La Grange Park, Illinois.

ASME, New York: Tables 2.3, 3.13; Figures 3.18 and 3.19.

American Vacuum Society, San Diego: Figure 10.2.

American Welding Society: Table 10.11; Figures 10.9, 10.10 and 10.11.

BAJ Vickers Co Ltd (M Donovan): Figure 10.3.

British Steel Corporation: Tables 3.19, 11.12; Figures 11.3, 11.4 and 11.5.

Butterworths, London (K E Thelning, *Steel and its Heat Treatment*): Figure 3.10.

Cambridge University Press (*Welding Process Technology*): Figures 11.8 and 11.9.

Carbide and Tool Journal: Figures 3.12 and 3.15.

Centre d'information du Chrome Dur, Paris: Tables 6.2 and 6.3; Figures 6.4 and 6.5.

Cobalt: Figure 6.6.

Deloro Stellite (Cabot Stellite Division), Swindon: Figures 11.3 and 11.11.

DoAll Co, Des Plaines, Illinois: Figure 3.22.

The Electrochemical Society Inc. These figures were originally presented at the Fall 1979 Meeting of the Electrochemical Society, Inc, held in Los Angeles, California: Tables 3.10, 6.4, 7.5 and 7.8; Figures 8.2 and 8.3.

Elsevier Sequoia SA: Tables 3.15, 6.6, 8.4, 9.6, 9.7, 9.8 and 10.9; Figures 3.16, 3.17, 6.7, 8.1, 9.3, 9.6 and 9.10.

Firth Brown Ltd, Sheffield: Figure 5.11.

Guhring Vertriebsgesellschaft, Albstadt: Table 3.14.

University of Aston in Birmingham (reproduced from *Heat Treatment of Metals*, the quarterly journal of the Heat Treatment Centre, The University of Aston in Birmingham, England): Figures 2.3, 3.1, 4.1, 4.2, 4.7, 5.4, 5.8, 5.15, 5.17, 5.20, 5.23, 5.24, 5.26, 5.27, 5.28 and 5.32; Tables 4.1, 4.2, 4.3, 5.4 and 5.9.

Ellis-Horwood Limited, Chichester, England (reproduced with permission from *Coatings and Surface Treatments for Corrosion and Wear Resistance*, by Strafford, Datta and Coogan, published by Ellis Horwood Limited, Chichester, 1984): Figure 5.36.

Industrial Heating: Figure 5.16.

Institute of Mechanical Engineers: Figures 1.3, 1.4 and 1.5.

Institute of Metal Finishing: Figures 2.4, 6.8 and 9.7; Table 7.1.

Institute of Metals, London: Figures 4.4, 4.5, 5.10, 5.12, 5.13, 5.14, 5.29 and 5.31.

Journal of Heat Treating, ASM, Metals Park, Ohio: Figures 5.18, 5.19, 5.22 and 5.25.

Materials and Design: Figures 3.23 and 3.24; Tables 5.2, 5.5 and 5.6.

Messer-Griesheim: Figure 11.7.

Metals Progress, ASM, Metals Park, Ohio. Figures 3.13 and 3.14.

Metallurgia: Figures 5.2 and 5.3.

Metallurgical Industries Inc, NJ: Figure 11.2.

Metallurgical Society of AIME (Reprinted with permission from *Journal of Metals*, Vol 34, No 9, 1982, a publication of The Metallurgical Society of AIME, Warrendale, Pennsylvania): Figures 9.11 and 9.12; Table 9.9.

Mitsui and Company Limited: Table 3.16; Figure 3.25.

Multi-Arc (Europe) Ltd: Figure 9.8.

National Association of Corrosion Engineers (NACE): Table 7.2.

NEI-International Combustion Ltd, Derby: Tables 3.21 and 3.22.

Nederlands Instituut voor Lastechniek: Figures 10.5 and 10.6.

PFD Limited: Table 8.5.

Plasma Technik A G: Figure 10.4.

Plating and Surface Finishing (P&SF is published by the American Electroplaters' Society, 12644 Research Pkwy, Orlando, FL 32826, USA): Tables 7.3, 7.6 and 7.7; Figure 10.8.

Plenum Publishing Co, NY: Figure 5.34.

Rolls-Royce Ltd: Figures 3.9 and (with Tribology International) 3.10.

Society for Advancement of Material and Process Engineering (SAMPE): Table 3.12.

SAE—Reprinted with permission ©1979, Society of Automotive Engineers Inc: Table 3.5, Figures 3.5, 3.7 and 3.8.

Societe Continentale Parker, Clichy: Tables 7.9 and 7.10, Figures 7.3 and 7.4.

Sulzer Bros Ltd, Winterthur, Switzerland: Table 8.3.

Toyota Research and Development Laboratories, Tokyo: Table 5.10; Figure 5.35.

Tribology International: Figures 1.2 and (with Rolls-Royce Ltd) 3.10; Tables 3.4 and 8.2.

TRW Valves Ltd: Table 3.6.

Vermont Tap and Die Co (Dr Henderer): Figure 3.21.

The Welding Institute: Table 9.4; Figures 9.1, 9.2, 11.1 and 11.14.

Department of Trade and Industry

Wear resistant surfaces in engineering

a guide to their production, properties and selection

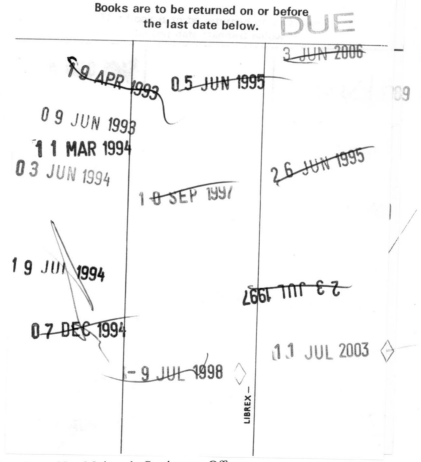
London   Her Majesty's Stationery Office

Contents

Chapter   *page*

Introduction

1 Types of Wear   2

    1. Introduction   2
    2. Adhesive wear..   2
    3. Abrasive wear..   4
    4. Fatigue wear   7
    5. Combined wear types   8
    6. Wear equations   9

2 Comparison of surface treatment processes   10

    1. Introduction   10
    2. Surface heat treatments   10
    3. Surface coatings   12
    4. Principal features of surface treatment processes   13
    5. Summary of factors in surface treatment selection   19

3 Applications   22

    1. Gears   22
    2. Automobile steering gear   25
    3. Cylinder liners and piston rings   26
    4. Poppet valves in internal combustion engines   28
    5. The aero gas turbine   31
    6. Synthetic fibre processing machines..   32
    7. Cemented carbide cutting tools   33
    8. High speed steel cutting tools   36
    9. Metal forming tools   40
   10. Rock, ore and earth engaging equipment   42
   11. Chute liners for coke and blast furnace sinter   44
   12. Pulverised fuel handling equipment..   45

4 Thermal hardening   48

    1. Introduction   48
    2. Induction surface hardening   48
    3. High-frequency resistance hardening   49
    4. Flame hardening   50
    5. Tungsten inert gas (TIG) hardening..   50
    6. Laser transformation hardening   51
    7. Electron beam transformation hardening   53

5 Thermochemical treatments   55

    1. Introduction   55
    2. Carburising   55
    3. Carbonitriding   60
    4. Nitriding..   61
    5. Nitrocarburising   65
    6. Comparison of thermochemical processes involving carbon and
       nitrogen   70

7. Boriding 75
8. Toyota Diffusion (or TD) process 76
9. Thermochemical treatments of non-ferrous alloys 77

6 Electrochemical treatments **81**

1. Introduction 81
2. Overview of properties and applications 81
3. Electroplating from aqueous solution 82
4. Brush plating 91
5. Electroplating from fused salts 92
6. Hard anodising 92

7 Chemical treatments **94**

1. Introduction 94
2. Electroless plating 94
3. Phosphating 99
4. Chromium oxide slurry coatings 101

8 Chemical vapour deposition **103**

1. Introduction 103
2. Deposition procedure 103
3. Deposition reactions 104
4. Coatings 104
5. Properties of coatings 106
6. Applications 107

9 Physical vapour deposition **110**

1. Introduction 110
2. Evaporation processes 111
3. Sputtering processes 113
4. Ion plating 116
5. Ion beam processes 120

10 Spraying processes **123**

1. Introduction 123
2. Processes 123
3. Bonding of coatings to substrates 137
4. Effect of coatings on fatigue strength 138
5. Wear resistance of coatings 138

11 Weld hardfacing **142**

1. Introduction 142
2. System selection 142
3. Hardfacing material 143
4. Processes and equipment 150
5. Costs 158
6. Current developments 160

Appendix

1 Wear test methods 161

2 Material specifications 164

3 Hardness measurements 166

Index 168

Introduction

Wear, together with corrosion and fatigue are the three principal processes limiting the useful life of engineering products. Examples of situations where wear is the predominant life-limiting factor are, the teeth in a mechanical digger which wear by contact with soil and rocks, the cylinder wall and piston rings in a heavy duty diesel engine, and a metal cutting tool which is worn by the material being machined.

The amount of wear which can be tolerated before there is a significant loss of performance varies greatly from component to component. For example, a digger tooth operates satisfactorily after several centimetres of wear whereas the performance of a metal cutting tool deteriorates when wear is a fraction of a millimetre. In some cases mild wear can be beneficial, as, during the 'running-in' of some equipment, it removes asperities from freshly machined surfaces which could be the origin of more severe wear when the equipment operates under more arduous conditions.

Methods of reducing wear have been under development throughout the industrial age but in the last few decades, with the growing realisation of the high cost of wear (estimates[1] range from $20–$100 billion pa in the US alone), the pace of development has accelerated. These developments include changes in design, improved lubrication, improved sealing and the use of more wear resistant materials. The latter may be bulk materials but they can also be surface layers on softer and tougher substrates.

A large number of wear resistant coatings and surface treatments are now available so that it is frequently difficult for the design engineer to select the optimum process for a given application. The aims of this book are to describe the main surface treatments available and to provide guidance in the process of selection.

The method that is suggested is:

1 Identify the type(s) of wear occurring in the component. This can usually be done from a knowledge of the conditions under which the component operates and by examination of the worn surface. In this way the basic requirements of the surface can be determined. (Chapter 1.)

2 From a knowledge of the salient features of various coating processes, eliminate those which are clearly unsuitable, eg inadequate hardness or bond strength or thickness limitation too high or too low. Produce a list of candidate coatings. (Chapter 2.)

3 Consult the case histories of the successful applications of the candidate coatings and by analogy reduce the candidate list as far as possible. (Chapter 3.)

4 Refer to the detailed description of the selected process(es) to confirm its suitability and to select the appropriate variant. (Chapters 4–11.)

The selection of the optimum treatment by such a process of elimination is not always possible and it may be necessary to test a number of treatments either in the laboratory or on the operating equipment.

In most cases there is more than one acceptable treatment in which case the cost and availability of the competing processes are major factors in determining final selection.

References

[1] PETERSON, M B, Introduction to Wear Control. *Wear Control Handbook*, ASME, NY 1980.

CHAPTER 1
Types of wear

1. Introduction

Wear is the loss of material from a surface caused by interaction with another surface or material. The main interactions are loads and motions producing adhesion, abrasion or fatigue, all of which can lead to the loss of wear fragments.

When adhesion occurs between two metals relative motion leads to material transfer which may produce a wear particle. Abrasive wear results from the ploughing, cutting or chipping action when a hard body slides or interacts with the surface. Where surface loads are applied repeatedly fatigue cracks may be formed which propagate to release wear fragments. These three principal wear types may be classified as shown in Table 1.1.

Table 1.1 Classification of wear processes

Adhesive wear	mild wear
	severe wear and scuffing
Abrasive wear	machining wear
	low stress sliding abrasion
	particle impact erosion
	three-body abrasion
	gouging abrasion
Fatigue wear	contact fatigue
	percussive wear
	cavitation erosion
	delamination wear
Combined wear types	fretting
	corrosive wear

Before a selection of materials or treatments to resist wear can be made, it is necessary to identify which wear processes are present, or are possible. An examination of worn surfaces, together with a knowledge of the mode of surface interaction, the surface properties and whether or not abrasive material is present, usually enables the type of wear to be identified. To aid this identification the following sections include descriptions of the appearance of surfaces worn by the different processes. Usually it is sufficient to examine surfaces visually or with the aid of a low power microscope. With the more detailed or difficult examinations the additional information provided by a scanning electron microscope, metallographic sectioning or wear debris analysis may be of assistance. There can be interaction between the wear processes, the factors which produce them and the functioning of the component. For example, the loss or movement of surface material, besides resulting in wear debris which is a potential abrasive, may lead to loss of fit, alignment or conjugacy, with resultant changes in loads and movement and subsequent wear damage. Examination of worn components should seek to identify the initial or rate controlling mechanism. In untried situations a consideration of all the possible wear processes is necessary to assess which may occur. Environmental factors which influence the wear mechanism, are reactions with the surface by gases, lubricants or corrosive compounds and frictional heating. Such effects cannot be generally categorised but are described in those cases where it may be significant.

2. Adhesive wear

Wear between two sliding engineering components can be of two main types: mild and severe. Machining wear and wear due to trapped abrasive grains can also occur but these are described later under Abrasive wear.

When two metal surfaces are loaded together, the opposing asperities make contact, deform and may weld together due to inter-atomic forces. With sliding motion present, and if the bond is stronger than the underlying material, the material shears or flows and work-hardening takes place. Material which is transferred from one surface to the other may later be removed as wear debris. This is termed adhesive wear which may be mild or severe.

(i) Mild wear

Many clean metals adhere or weld together strongly, whereas non-metallic materials or phases have weak adhesion. Under normal atmospheric conditions, all metallic surfaces are covered with a layer of adsorbed gases, vapours or chemical reaction products (usually oxides) so that the adhesion between the surfaces is low and any junctions have low shear strength. The oxide layer is generally very thin and is easily penetrated by work-hardened asperities to expose the clean metal which will have a strong tendency to adhere to the opposing surface if it has also had regions of oxide removed. Re-oxidation of the exposed metal commences immediately, except in vacuum or inert atmospheres.

Mild wear results when the wearing conditions are not severe enough to remove the oxide film faster than it can be re-formed. An oxide film of low adhesion is maintained and fine oxidised debris and a smooth surface result. This type of wear occurs in most moderately loaded sliding applications and is often termed oxidational wear. Mild wear also results when sliding materials have low adhesion, irrespective of whether oxide films are formed, eg with hardened steel, non-metallic materials or themochemically treated surfaces. Surface hardness is a principal factor affecting mild wear but microstructural factors are also important. Adhesive wear is reduced if the structure is discontinuous, eg the lamellar pearlitic structure in some steels.

(ii) Severe wear and scuffing

Severe wear may result at higher speeds and/or loads, or under other conditions in which a stable oxide film is not maintained at the surfaces. It is characterised by non-oxidised torn surfaces and coarse metallic debris. The wear rate may be an order of magnitude greater than that for mild wear.

As load is increased the rate of both the mild and severe types of wear increases. Also, as shown in Fig. 1.1., a transition may occur from mild to severe wear at a critical value of load or speed. At even higher loads and/or speeds, or when higher temperatures are otherwise developed, the oxide film may again predominate and effect a reverse transition from severe to mild wear. Indeed those materials which have good oxidation and corrosion resistance tend to have poor resistance to adhesive wear. However, excessive oxidation is undesirable because a large amount of free oxide will lead to abrasive wear. Many engineering components undergo a period of severe wear during running-in before mild wear is established.

Scuffing is a form of severe wear with tearing and plastic flow which takes place when the fluid or solid lubricant film separating sliding metallic surfaces breaks down at high temperatures. In some situations there is no distinction between scuffing and severe wear but generally scuffing is associated with high duty operating conditions in which severe wear causes increasing temperatures and accelerating damage. It is mainly a problem which occurs during the running-in of components such as gears, cams, piston rings and cylinders which experience high concentrated loads, high sliding velocities or a combination of both. Scuffing varies in degree from 'slight', which is characterised by a dull surface, to 'heavy', which shows more obvious signs of metallic tearing, in the direction of sliding. If allowed to continue, scuffing can be a destructive process leading to seizure or component failure due to increased levels of stress or vibration.

Severe wear and scuffing are best resisted by the use of materials of low adhesion (eg dissimilar materials of low mutual solubility). Increasing the hardness of steels is beneficial and with hardnesses over 700 HV* adhesion is minimal except under arduous conditions in which frictional heating softens the surface. Microstructural factors must also be considered when selecting materials to prevent severe wear and scuffing. Fig. 1.2 shows wear values[1] for steel discs of three carbon contents, lubricated with SAE 30 oil, sliding in the Amsler machine.† Although each steel is in the normalised (ie soft) condition, and the hardnesses are not greatly different, the higher the carbon content the greater is the resistance to wear.

---

*See Appendix 3 for discussion of hardness values.
†See Appendix 1 for description of wear test methods.

Fig. 1.1   Influence of load and speed on sliding wear.

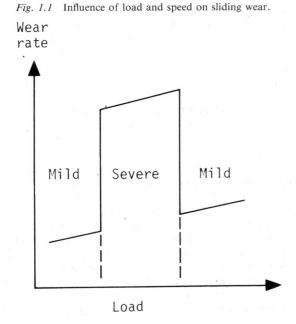

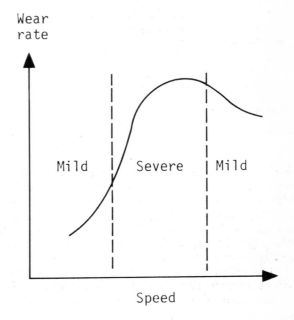

Microstructurally the three steels are quite different. The low carbon steel being ferritic and the high carbon steel consisting almost entirely of pearlite. The superiority of pearlite is due to its lamellar structure of hard cementite (700 HV) and ferrite. Martensitic structures, despite their high hardnesses, do not have outstanding scuffing resistance as this structure can transform to ferrite and pearlite at local hot-spots. Thermochemical treatments such as nitrocarburising and nitriding which produce non-metallic nitrides of low adhesion are good anti-scuffing treatments. As scuffing may be only a running-in problem it can often be prevented by the use of short life treatments, such as soft metal platings and phosphating, which give protection until an improved surface finish is developed.

3. Abrasive wear

Abrasive wear is the removal of material from a surface by harder material impinging on or moving along the surface under load. The hard material indents the surface and, depending on the properties of the materials and the type of motion or loading, may remove material by various mechanisms, eg cutting, ploughing, chipping or by fatigue cracking.

(i) Machining wear

When a surface having hard asperities slides on a softer surface, machining wear may result. Wear debris like machined swarf is produced by the cutting or ploughing action of the hard asperities through the softer material. Surfaces resulting from machining wear have a scored and grooved appearance. A typical example is with badly finished worm-gears in which a coarsely ground hard steel surface slides on a softer bronze surface producing high rates of machining wear.

Therefore, when a hard surface slides on a soft surface, a good surface finish (eg better than 0.2 μm CLA*) is desirable. A soft metal plating applied to the harder surface can be beneficial in limiting asperity penetration of the softer surface by the presence of load carrying material in the valleys between the asperities.

Alternatively, hard abrasive grains of work hardened wear debris can become embedded in one or both surfaces of a sliding pair to form sharp cutting edges. If the two surfaces are of different hardnesses the particles will tend to become embedded in the softer surface, causing wear of the harder surface.

As machining wear can also be considered a form of low stress abrasion, the measures required to minimise it are similar. Frequently the most easily replaced

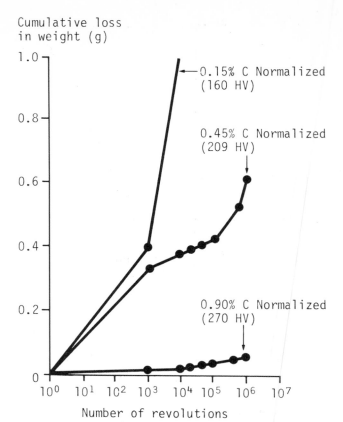

Fig. 1.2 Influence of carbon content on wear rate of steel.

member of the sliding pair is made from material having lesser wear resistance than the other member.

(ii) Low-stress sliding abrasion

When abrasive material slides over a surface at relatively light loads without significant impact, the wear produced by the cutting or ploughing action is termed low-stress abrasion. This type of wear is common in materials' handling equipment, eg chutes or components, such as ploughshares, in rock-free soil. It also includes wear by abrasive matter in fibre and thread or by any semi-solid medium which can apply a force between the abrasive particles and the surface. Generally the worn surface has a scratched appearance but, if the abrasives are fine, the surface is polished.

Hardness is the most important property in resisting low-stress abrasion. For pure and annealed metals, worn by a harder abrasive, the abrasive wear resistance, which is the inverse of the abrasive wear rate, increases linearly with metal hardness[2] as shown in Fig. 1.3.

With materials of more complex microstructure (eg hardened steels) the increase in abrasion resistance with hardness, as related to the annealed state, is lower than that obtained for pure or annealed metals of equivalent hardness (Fig. 1.4). Prior work hardening of a material has little influence on the resistance to low stress abrasion as shown in Fig. 1.5. The influence of abrasive hardness on the wear rate of various materials is shown schematically in Fig. 1.6.

---

*CLA—Centre Line Average—A measure of surface roughness defined as the average value of the departure of the profile from the centre line throughout the sampling length. The equivalent terms roughness average (Ra) and arithmetic average (AA) are also used. (See for example 'Exploring Surface Texture'— H. Dagnall—published by Rank Taylor Hobson.)

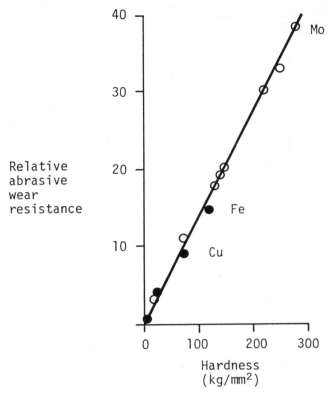

Fig. 1.3   Wear resistance of commercially pure metals.[2]

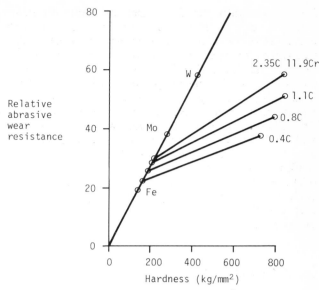

Fig. 1.4   Wear resistance of heat treated steels.[2]

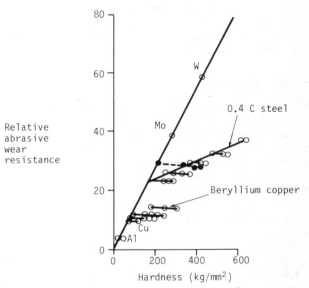

Fig. 1.5   Wear resistance of cold worked metals.[2]

When the hardness of the materials is greater than that of the mineral the wear rates are low. Very little increase in wear resistance is obtained by increasing the material hardness to more than 1.3 times the mineral hardness. A steep rise in the wear rate occurs when the mineral is harder than the metal but this levels off with further increases in mineral hardness. Surface hardness may be chosen either to exceed the abrasive hardness or to obtain an increase in wear resistance as shown in Fig. 1.6. A wide range of materials is used to resist low-stress abrasion, the choice being determined by the hardness of the abrasive and the prevailing operating and environmental conditions.

(iii) Particle impact erosion

This is a type of wear which results when a stream of liquid or gas containing sharp particles impinges on a surface, eg in pipelines carrying slurries and during shot blasting.

When sharp particles strike the surface at an acute angle, material is removed by a micro-cutting action and the surface is scratched in the direction of fluid flow in the localised areas of impingement. High surface hardness is used to resist this form of erosion and brittle materials may be used since fracture is unlikely at low impact angles unless the particle velocity is high.

At high angles of impingement the repeated impacts result in wear particles by deformation and fatigue in ductile materials or by microcracking in brittle materials. A deformed or roughened surface is evident with some surface pitting which may be on a large scale if the impact velocity is high. The ability of a material to resist high impact angle erosive wear depends on the amount of energy it can absorb before fracture, as indicated by the area under its stress strain curve. Thus, tough ductile materials are superior to hard brittle materials when the impact angle is high.

Both the cutting and the fracture types of wear will normally occur simultaneously and the relative amount of each varies according to the angle of impact. The influence of impact angle on wear rate is shown in Fig. 1.7. The deformation and brittle fracture types of wear can also be produced by high velocity liquid droplets, eg at steam turbine blade leading edges.

(iv) Three-body abrasion

When loose abrasive particles of small size are present between two sliding or rolling surfaces, the wear is termed three-body abrasion (if the abrasives are large, as in roll crushing, this three-body definition is not generally applied).

In a three-body situation the resulting wear mechanism depends on factors such as the hardness, fracture strength and size of the particles, the hardness and roughness of the surfaces and the type of motion imposed. When hard particles are present between sliding soft surfaces, as with grit laden journal bearings, there will be a tendency for the particles to become permanently embedded and the situation will revert to two-body abrasion, eg machining wear. If the abrasives are not significantly harder than the sliding surfaces they only become temporarily embedded in the surface and, being free to roll, the wear rate is an order of magnitude lower than with fixed particles. With abrasive particles which are smaller than the scale of the surface roughness little wear is caused by the particles.

Grinding abrasion, often termed high stress abrasion, is a particular type of three-body abrasion in which the abrasive grains are crushed. This type of abrasion takes place principally at the surfaces employed to grind small sized rocks, eg in grinding mills, but it can also occur with sliding or rolling engineering surfaces operating in gritty conditions if the strength of the abrasive is exceeded and the abrasive is fractured and sharp angular fragments are produced. The worn surfaces show short deep scratches, grooves and indentations. There is little opportunity for the abrasive to cut or roll before fracture occurs; it is therefore likely that the wear of the abraded surface arises partly from cutting and partly from successive indentation by the grains during fracture. With ductile materials local deformation probably causes fatigue damage with resulting loss of material, whereas with brittle materials microcracking and chipping occur.

Materials having a combination of hardness and toughness to resist indentation and cracking respectively are the most suitable materials to resist grinding abrasion.

(v) Gouging abrasion

Gouging abrasion describes the process where relatively large particles of material are torn from a surface by interaction with coarse rocks or abrasive

Fig. 1.6 Wear rate of materials v. mineral hardness.[3]

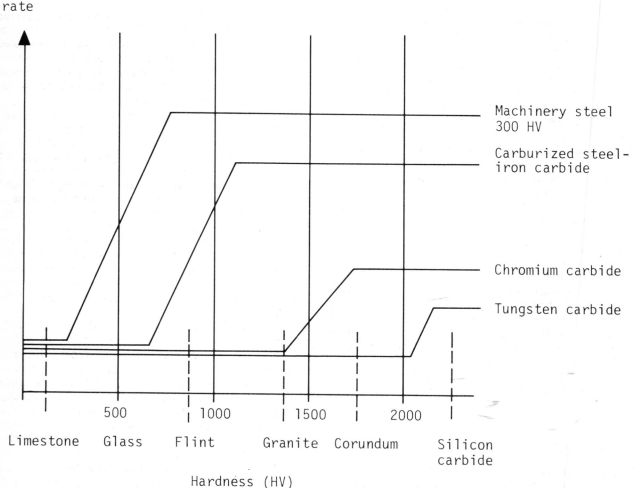

Wear rate

Machinery steel 300 HV

Carburized steel-iron carbide

Chromium carbide

Tungsten carbide

500   1000   1500   2000

Limestone   Glass   Flint   Granite   Corundum   Silicon carbide

Hardness (HV)

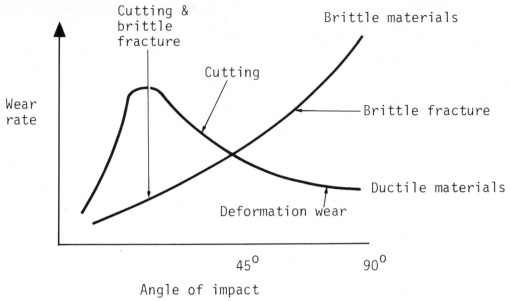

Fig. 1.7 Effect of impact angle on erosive wear.

material. Whilst there may be two metal surfaces involved, the second surface, which produces the load on the abrasive fragment, does not directly interact with the first. Gouging abrasion may occur at relatively low speeds, eg in excavator buckets, or at high speeds, eg in impact pulveriser hammers or roll crushers. The surfaces generally acquire a deeply gouged, scratched and possibly pitted appearance. The mechanism of metal removal may involve deformation, cutting or chipping, depending on the sharpness of the abrasive and the properties of the material but gouging implies metal removal by grooving on a scale larger than the scale of the microstructure of the abraded material. It is this which distinguishes it from low stress abrasion though the latter is usually also present. As gouging is often accompanied by severe impact the selection of the material is a compromise between abrasion resistance (hardness), and impact strength or toughness. The relative importance of the two properties will be dictated by the service conditions.

There is a long history of austenitic manganese steel being used for components or facings, to resist gouging abrasion with impact because of the material's combination of high toughness and hardness after becoming work hardened. However, harder materials with better abrasion resistance may be used in less severe impact conditions. A variety of materials, mainly steels, irons or tungsten carbide composites, are available with various ratios of hardness to toughness. Because of the relatively large depth to which abrasives may gouge out material, thin cases or layers are ineffective in reducing this type of wear.

4. Fatigue wear

Surfaces can wear by several processes which are initiated and controlled by a fatigue mechanism.

These processes include contact fatigue wear, percussive wear, cavitation erosion and delamination wear.

(i) Contact fatigue wear

Surfaces can wear by loss of fragments separated from the surface by fatigue cracks. When a component is subjected to repeated or fluctuating concentrated contact loads it may, for a considerable time, function without apparant distress but eventually a crack forms even though the applied stresses are well below the bulk material's ultimate strength.

The cracks form at or beneath the surface and propagate in a direction dependent on the sub-surface stress distribution and stress-raising discontinuities. With concentrated loads (eg in gears, cams and rollers) the cracks often originate at the surface and propagate into the material along a curved path leading back to the surface. The resulting damage is termed pitting as smooth bottomed cavities or pits are formed. The number of stress cycles required to initiate pitting decreases with increasing stress level and data is typically presented as stress-v-cycles (S/N) curves. Pitting, once initiated, is progressive unless arrested by increased material strength due to work hardening or by reduced localised stresses after bedding-in.

Hardened (eg carburised, nitrided, and thermally hardened) layers are widely used to resist pitting. Besides having increased fatigue strength the case usually has beneficial residual compressive stresses. Care is needed in selecting the appropriate case depth in relation to the applied stress levels and the strength of the underlying material as, with loaded counterformal surfaces, sub-surface shear stresses reach a maximum at some depth below the surface. Sub-surface fatigue may be initiated if the shear stresses are sufficiently high. Also, if the material below the case deforms excessively under load, fatigue cracks may

7

be initiated at the interface between the substrate and the hardened case. Cracks will propagate readily along a case/subsurface boundary if the hardness gradient is severe or if the bond, as with overlay coatings, is poor. Where large areas of case are removed the damage is termed spalling or case-exfoliation.

Surface fatigue may also be initiated by alternating stresses produced by the expansions and contractions of a material subjected to temperature gradient fluctuations. This is termed thermal fatigue and, in a wear context, is mainly associated with non-ductile materials having poor thermal conductivity, although if the temperature gradients are severe other materials can suffer from this type of damage.

(ii) Percussive wear

Percussive wear may result when one surface is stressed by impact with another. If, due to heavy impact or local stress concentrations, the impact stress exceeds the ultimate strength of the material, cracking or fracture may result from a single impact. When the impact stress is merely above the yield strength of the material, successive impacts produce continuing permanent deformation which eventually reaches the maximum deformation which the material can sustain before cracking occurs. Because of work hardening during plastic deformation, the yield strength of the material may increase to a value above the applied stresses, so arresting further deformation and preventing cracking and wear. Where the stresses are below the yield strength of the material a third type of wear, surface fatigue, may occur due to repeated impact stresses.

For light impact duties it is often possible to use materials of high elastic strength to prevent deformation cracking. In situations of heavier impact, materials having toughness and work hardening ability are necessary. Thin hard surface layers are particularly liable to cracking if the substrate is too soft and deforms plastically under localised impact stresses. However, a softer substrate can be beneficial in absorbing impact loads but in this case the hard surface material, deposited for example by welding, must be thick enough to distribute any local surface stresses to a large area of substrate. In many situations abrasive wear also has to be resisted and the choice of material is a compromise. Hard inserts are frequently employed to resist both impact and abrasion. Economic considerations sometimes allow cheaper, less wear resistant materials to be used if part replacement is convenient. Where the impact is of a vibratory nature fretting may also occur and require consideration in the choice of material.

(iii) Cavitation erosion

Cavitation in liquids is the formation of bubbles or pockets where the liquid pressure falls to its vapour pressure and boiling occurs. When the liquid carries the bubbles into higher pressure regions they collapse and if a material surface forms part of the bubble boundary the surface is subjected to impact by liquid. This can lead to a type of surface wear known as cavitation erosion which occurs in hydraulic components and marine propellers. Cavitation eroded surfaces show surface pitting but often this form of wear is accompanied by some corrosion by the liquid.

Surface material selection is generally dictated by considerations such as corrosion resistance but where possible it is useful to have a high ultimate resilience characteristic:

$$\text{ultimate resilience} \atop \text{characteristic} = \tfrac{1}{2} \frac{(\text{tensile strength})^2}{\text{elastic modulus}}$$

Cast cobalt alloys (particularly Stellites 12 and 4) are very resistant to severe cavitation erosion, as are high strength titanium, ferrous, copper and nickel alloys. Hard chromium plate or other corrosion resistant coatings are also useful.

(iv) Delamination wear

Delamination wear is characterised by the release of small sheets or platelets of material from a wearing surface. The process is initiated by sub-surface deformation and fatigue cracking due to the repeated normal and traction loads imposed by the sliding counterface. Research on this type of wear is in its infancy and there is disagreement on whether delamination is a basic mechanism of wear which manifests itself in other forms of wear such as adhesive and abrasive wear or if delamination is to be regarded as an observable type of wear whose basic mechanism is fatigue. The rate of delamination wear is determined by microstructural features such as density and type of inclusions and dispersed phases. Hard single phase materials or materials with coherent second phase particles will resist this type of wear. If the material contains incoherent second phase particles, inclusions or porosity, delamination wear is more likely to occur.

5. Combined wear types

In many wear situations there is a possibility that two or more wear processes may occur simultaneously. In some cases the combination of processes produces a wear type which is quite different from either of the basic processes or mechanisms. The most well known such cases are fretting and corrosive wear.

(i) Fretting

Fretting is a form of surface wear which occurs when two contacting surfaces undergo very small amplitude oscillatory or vibrational slip. It can also be associated with conditions of vibratory impact in which percussive and fatigue wear may be present. Fretting is characterised by the production of a finely oxidised debris which often flows out of the contact area. The debris on ferrous metals has a cocoa-like appearance and consists of ferric oxide. The fretted surfaces are generally pitted; the pits being shallow, fully oxidised and surrounded by the debris.

Fretting is a particularly serious form of wear which, when occurring at surfaces with intended relative motion (eg bearings and linkages), may lead to further damage or seizure. It can also arise where no relative motion is intended, eg with press fits where loss of fit results. Surface damage caused by fretting can also lower the fatigue strength due to the development of fatigue cracks.

A widely accepted theory of fretting is that adhesive wear produces very small loose particles which become oxidised and, as they are trapped between the oscillating surfaces, cause abrasive wear at a higher rate. However, the mechanism is not fully understood and an alternative theory is that surface fatigue aided by the reaction of the surface with oxygen is responsible.

Although fretting is best prevented by eliminating the oscillations, it can often be reduced by choice of materials, treatments or coatings. Dissimilar materials are generally preferred but if this is not possible the use of a soft metal coating on one component is beneficial. Harder or non-metallic surfaces as produced by nitriding, chromium plating, flame or plasma spray deposition may also be useful if adequate fatigue resistance or bonding can be achieved. Aluminium and titanium may be anodised to resist fretting.

(ii) Corrosive wear

Corrosive wear is defined here as a wear process which is at least partly chemical and in which the chemical effect is not beneficial. Many metals are resistant to progressive reaction with mildly reactive media by virtue of oxide films or passive reacted layers. In a wearing situation, these films or layers may be removed by mechanical interaction so that bare metal is continuously exposed to the medium, eg chains operating in hot steamy conditions. In such situations, media not normally corrosive to the protected metal become so and this may accelerate the rate of loss of material to a value greater than that which would be lost by either the corrosion or the wear process alone.

Corrosion and corrosive wear increase with increasing temperatures. At high temperatures (say, above 500°C) the process is mainly one of gross oxidation, and frequently high chromium alloys are used because of their ability to form protective chromium oxide layers. In any wear situation involving fluid flow (eg impact by fluids and cavitation erosion) corrosion may also be a problem. It is impossible to generalise on the appearance of corrosively worn surfaces because such a large number of situations are possible that identification is a matter of judgement.

Selection of surfaces to resist corrosive wear is usually made on the basis of preventing the corrosive effect.

6. Wear equations

Most published wear studies refer to the Rabinowicz[4] and Archard[5] wear equation:

$$V = \frac{k.L.S.}{H} \qquad \text{equ (1)}$$

Where V is the wear volume, k is the wear coefficient. L is the load, S is the sliding distance and H the hardness of the wearing surface. This equation has been applied to mild sliding wear, adhesive wear and abrasive wear. The usefulness of the wear coefficient, k, is limited as it may not be constant for a given material and may change with applied load, sliding velocity, temperature and other tribo-conditions.[6] There is an increasing acceptance that as k and the relevant hardness depend on material microstructure,[7] H and k should not appear independently in the equation[8] and the wear constant K in the following equation must be experimentally determined:

$$V = K.L.S. \qquad \text{equ. (2)}$$

However, some authorities[9] suggest that wear may be predicted from published tables of wear coefficients. Such predictions can only be accurate if the operating conditions are close to those under which the coefficients were determined. There is at present insufficient wear data available to permit wear to be calculated for most engineering applications.

References

[1] GREGORY, J C, Thermal and chemico-thermal treatments of ferrous materials to reduce wear. *Tribology*, Vol 3, No. 2, p 73 May 1970.

[2] KRUSCHOV, M N, Resistance of metals to wear by abrasion as related to hardness. *Proc. Conf. on Lubrication and Wear, Inst. Mech. Eng. London*, p 655, 1957.

[3] FARMER, H N, Selection and performance of hard facing alloys. *Met. Eng. Quart.,* Nov. 1975.

[4] RABINOWICZ, E, *Friction and wear of materials.* Wiley, New York, 1966.

[5] ARCHARD, J F, Contact and rubbing of flat surfaces, *J. Appl. Phys.,* 24 p 981, 1953.

[6] VERBEEK, H J, Tribological systems and wear factors, *Wear,* 56, p 81 1979.

[7] SAKA, N, Effect of microstructure on friction and wear of metals, pp 135–170, in *Proc. Int. Conf., Fundamentals of Tribology,* Cambridge, Mass. USA, 1978. ed. Suh, N P and Saka, N, 1980.

[8] JAHANMIR, S. On the wear mechanisms and the wear equations, pp 455–567, in *Proc. Int. Conf., Fundamentals of Tribology,* Cambridge, Mass. USA, 1978. Ed. Suh N P, and Saka, N, 1980.

[9] RABINOWICZ, E, Wear coefficients—metals, *Wear Control Handbook,* ed Peterson, M B and Winer, W O. ASME, NY 1980.

Additional reading

Tribology Handbook ed Neale, M J. Butterworths, London, 1973.

BUCKLEY, D J. *Surface effects in adhesion, friction, wear and lubrication*, Tribology Series Vol. 5, Elsevier 1981.

CZICHOS, H. *Tribology—A Systems approach to the Science and Technology of Friction Lubrication and Wear.* Tribology Series, Vol. 1, Elsevier, 1978.

WATERMAN, N A. *Fulmer materials optimiser*, Vol 1, 1st edn. Fulmer Research Institute, Stoke Poges. 1974.

CHAPTER 2

Comparison of surface treatment processes

1. Introduction

In the previous chapter an indication has been given of the material properties required to resist the various forms of wear. These properties can be obtained in bulk material or in layers produced on the surface of different substrate materials. Most wear resistant materials are hard and have low values of toughness (the ability to absorb impact energy). They are also more difficult to process than the metals used in engineering and consequently it is usually only components of simple shape which are made from them. The advantages of using surface layers are that a tougher substrate can be used and components of complex shape can be made more cheaply.

Methods of producing a hard surface on metal components can be divided into two broad categories; those in which the hardened surface is produced by a heat treatment, usually accompanied by or preceded by the diffusion of elements, eg carbon or nitrogen, into the surface and those in which a layer of hard material is formed or deposited on the surface. The principal processes in these two categories are listed in Fig. 2.1. Most processes have several variants which are described in the appropriate chapter of this book.

Although some surface heat treatments are applicable to non-ferrous alloys they are used mainly on steels. The surface layers produced are integral with the substrate and there is usually a gradation in composition and structure from the surface to the substrate. Surface coating processes such as electrochemical and chemical treatments, physical vapour deposition and spraying in which the substrate is relatively cold during deposition can be used on a wide range of substrates. The high temperatures occurring during welding and chemical vapour deposition make these processes unsuitable for some substrates and restrict the choice of coating composition which can be applied to others.

2. Surface heat treatments

(i) Thermal hardening

In thermal hardening a hard layer is produced on plain carbon and low alloy steels of medium carbon content (0.3–0.6%) by rapid heating of the surface followed by water or oil quenching to form martensite. The process is cheap but the compositions of the steels on which it can be used are limited. It has the advantage that selected areas can readily be heat-treated. The depth of hardening can be varied by the method of heating and is normally in the 1–10 mm range.

(ii) Carburising and carbonitriding

Carburising is a process in which carbon (up to 0.8%) is diffused into the surface of a steel which is subsequently hardened and tempered. It is carried out in the temperature range 850–950°C to obtain reasonable carbon diffusion rates. The process is applicable to a wider range of steel compositions than is thermal hardening and therefore a wider range of properties of both the surface layer and the core can be obtained. Some distortion occurs during processing and subsequent grinding is frequently necessary. Case depths up to several millimetres are readily obtained.

Carbonitriding is a variation of carburising in which nitrogen (up to 0.5%) is diffused into the steel along with carbon. Because nitrogen lowers the ferrite-austenite transformation temperature and also increases hardenability, a lower treatment temperature and a lower quenching rate can be employed both of which reduce the amount of distortion compared with carburising.

(iii) Nitriding and nitrocarburising

In nitriding, nitrogen is introduced to the steel surface by heating to 500–525°C in an atmosphere of ammonia or exposing it to a low pressure nitrogen-hydrogen atmosphere in an electric glow discharge. To obtain high surface hardnesses (>750 HV) nitride forming elements such as Al, Cr, Mo and V must be present in the steel. As the process is carried out at a low temperature and no subsequent heat treatment is required there is little distortion of the components. Moreover, the hardness developed is temper resistant up to at least 500°C. However, because of the low temperature of treatment, process times are much longer than in carburising for a given depth of surface layer. Layer depths are therefore usually not greater than 0.7 mm.

Nitrocarburising is a variation of nitriding and is usually carried out at 570°C in either a salt bath containing a mixture of sodium cyanide and cyanate or in a gas mixture containing ammonia and a carburising gas. A treatment time of two hours gives a 20 μm layer of a carbonitride phase which has good resistance to adhesive wear.

There are a number of variations of the basic nitrocarburising processes which are known by proprietary names, eg Tufftride, Tenifer and Nitrotec. In some variations sulphides are added to the salt bath

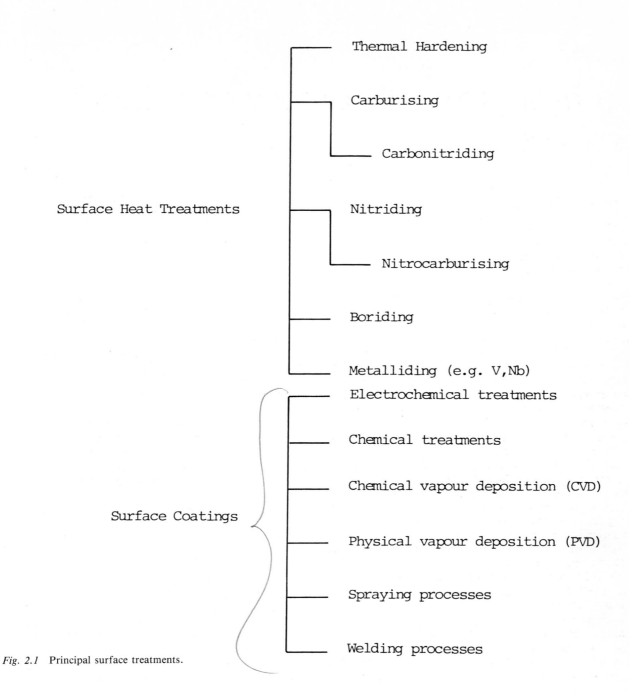

Fig. 2.1 Principal surface treatments.

Surface Heat Treatments
— Thermal Hardening
— Carburising
— Carbonitriding
— Nitriding
— Nitrocarburising
— Boriding
— Metalliding (e.g. V, Nb)

Surface Coatings
— Electrochemical treatments
— Chemical treatments
— Chemical vapour deposition (CVD)
— Physical vapour deposition (PVD)
— Spraying processes
— Welding processes

and these process variations are known as Sursulf and Sulfinuz.

Another salt bath treatment, Sulf BT, in which only sulphur is added to the surface and which is carried out at 200°C also provides good resistance to scuffing.

In addition to the above long established and widely used treatments[1] a number of other processes have recently been developed and are finding increasing application. The two more important of these are boriding and metalliding.

(iv) Boriding (boronising)

In this process boron is diffused into the surface of plain carbon and low alloy steels at approximately 950°C to form a layer of iron borides about 100 μm thick with a hardness in the range 1800–2100 HV. The process can also be used on cobalt, nickel and titanium alloys.[2]

(v) Metalliding

The most important process in this category is one developed by the Toyota company of Japan and known as the Toyota Diffusion (or TD) Process[3] in which carbide forming elements such as vanadium and niobium are diffused into steels from a salt bath at about 1000°C. Carbide layers (5–12 μm thick) of very high hardness ($\sim$3000HV) are produced.

The term metalliding is also used to describe the

11

similar process in which coatings are electroplated from fused salts.

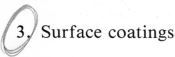 3. Surface coatings

(i) Electrochemical treatments

Electrochemical coatings are produced by electrolysis of an aqueous solution of a salt of the coating metal, the component to be coated being the cathode. For wear resistance, chromium is the coating most widely used as it combines high hardness ($\sim$1000 HV) with corrosion resistance and a low value of friction against steels. Chromium coatings are limited to a thickness of 0.5 mm because of internal stresses. Thicker coatings, up to several millimetres, of nickel can be obtained but as the deposit is relatively soft ($\sim$250 HV) is seldom used for wear resistance. Hard particles of carbides (eg SiC, Cr_2C_3) and oxides (eg Al_2O_3) can be incorporated in electrochemical coatings of nickel and cobalt during plating to give a coating hardness of 600 HV.

(ii) Chemical treatments

Chemical coatings are produced by the immersion of the component in a solution of a salt of the coating metal, no impressed current being used. Coatings of nickel phosphorus and nickel boron alloys are produced by the reduction of a nickel salt by sodium hypophosphite or sodium borohydride respectively. Coatings after heat treatment have a hardness of up to 1000 HV and have very good adhesive wear resistance. Electroless nickel coatings are more expensive than electrochemical chromium coatings but they have superior corrosion resistance and uniform thickness can be obtained on complex shaped components. Hard particles, eg of SiC, can also be incorporated in the nickel during deposition to give a composite coating with a hardness of $\sim$1300 HV. Thin (0.01 mm) coatings of metal phosphates are also formed chemically on steel components to provide surfaces of low friction and resistance to adhesive wear.

(iii) Chemical vapour deposition

Chemical vapour deposition (CVD) is a process whereby compounds are reacted in the gas phase to form a dense layer on a heated substrate. The most widely used wear resistant materials deposited by CVD are titanium carbide and titanium nitride and with these materials coating thicknesses are limited to about 10 μm by interfacial stresses. Temperatures in the range 800–1000°C are required for deposition and at these temperatures thermal distortion and chemical reaction between coating and substrate limit the choice of substrate. In wear resistant applications, only cemented carbides and some tool steels are used and in these cases the reaction which takes place results in very high bond strengths. Compared with PVD processes CVD has much better 'throwing power' ie the ability to coat complex shaped components with a coating of uniform thickness.

(iv) Physical vapour deposition

Physical vapour deposition (PVD) processes are performed at sub-atmospheric pressures, the coating atmosphere being generated by thermal evaporation or electrical sputtering of a suitable source material, possibly with additions of a reactive gas. Coating rates of titanium nitride, the most widely used wear resistant material deposited by PVD, are limited to a few microns per hour and coating thicknesses are usually in the 1–10 μm range. Coatings are dense and, with modern techniques, well-adhered to the substrate. Substrate temperatures can be maintained below 500°C.

(v) Spraying processes

A number of processes have been evolved in which particles of the coating material are heated to a molten or plastic state and projected at the substrate which is relatively cold ($<$200°C). Coating density and strength of bonding to the substrate increase with projection velocity. This is about 100 m/s in a simple combustion gun; 500 m/s in a plasma gun in which gases, usually argon and hydrogen, are heated in an electric arc to 15,000°C; and 800 m/s in a gun in which metered quantities of acetylene and air are detonated. Most metals and ceramics can be sprayed onto a wide range of metallic substrates. All sprayed coatings are porous to some extent, the porosity varying from about 20% for combustion gun to less than 1% for detonation gun coatings. The bond between the coating and the substrate is mainly mechanical and is usually considerably less strong than those obtained by other coating processes.

(vi) Welding processes

All of the various welding methods can be used to deposit wear resistant coatings (hardfacing). Coating materials range from low alloy steels to tungsten carbide composites. Rates of deposition are high and good bond strengths are obtained because of alloying at the coating-substrate interface. There is no easily defined upper limit to coating thickness although there is a tendency for some of the harder materials to crack if deposited too thickly. These cracks are not usually harmful since they are perpendicular to the surface and do not usually reach the interface. It is, however, impracticable to produce coatings less than 2 or 3 mms thick.

(vii) Solid tiles

In some applications, where a large amount of wear is tolerable, solid tiles, attached to the substrate mechanically, by adhesives, by cements or by brazing, are more economical than hardfacing. Materials used include alumina, fused basalt, 13% Mn steel, high chromium cast iron and tungsten carbide-cobalt cermets.

4. Principal features of surface treatment processes

The more important features of surface treatment processes which have to be considered in surface treatment selection are discussed below. Those which are materials related are summarised in Table 2.1 and those that are more related to the surface treatment process are summarised in Table 2.2. However, this division is only for convenience of discussion as the properties of the materials are influenced by the treatment process.

Table 2.1 Properties of the principal materials used to provide wear resistance

Wear resistant material	Hardness, HV	Maximum service temp °C	Corrosion resistance		Applicable coating methods
			Overall rating	Remarks	
Metals hardened by dispersed oxides					
eg Mo	390	250	Moderate	Attacked by mineral acids	Flame spraying
13% Cr steel	330	600	Good	Attacked by mineral acids	Flame spraying
Intermetallics					
eg NiAl	250–350	850	Poor	Attacked by acids and alkalis	Flame and plasma spraying
CoCrMo (Laves Phase)	1100	1000	Very good	Slowly attacked by hot HCl	Plasma spraying, welding
Hardened steels					
eg 12% Mn steel	150–400 (work hardens)	150–200	Poor	Attacked by acids and alkalis	Welding
Martensitic steels	300–850	150–200	Poor-moderate	Depends on alloy content	Welding, thermal and thermochemical treatment
Nitrided steels	800–1200	500	Good-moderate	Better than untreated steel	Thermochemical treatment
Cast irons					
eg Martensitic irons	400–600 (1350)	200–250	Poor-moderate	Depends on alloy content	Welding
High chromium irons	400–600 (1700)	1000	Good	Attacked by oxidising acids	Welding
Ni,Co,Cr alloys					
Cobalt alloys	300–700	850	Very good	Slowly attacked by HCl	Welding, plasma spraying
Nickel alloys	350–700	850	Very good	Slowly attacked by HCl and HNO_3 in some concentrations	Flame spray plus fusion, welding
Nickel alloys (+ tungsten carbide)	350–700 (2400)	850	Very good	Attacked by hot HCl and HNO_3 in some concentrations	Flame spray plus fusion, welding
Nickel phosphide; nickel boride	850–950	300	Very good	Slow attack by oxidising acids Boride more susceptible	Chemical treatments
Chromium metal	850–1000	350	Very good	Slow attack by mineral acids	Electrodeposition
Carbide-metal cermets					
eg WC-Co	1300–1600 (2400)	550	Good	Co slowly attacked by mineral acids	Detonation gun (D-gun) and plasma spraying
Cr_3C_2-Ni–Cr	1100 (2200)	820	Very good	Only attacked by hot HCl	D-gun and plasma spraying
Cr_3C_2-Co	450–500 (2200)	800	Good	Co attacked by mineral acids	Electrodeposition
WC-steel	500 (2400)	300	Poor-moderate	Depends on steel matrix	Welding
Oxides					
eg Al_2O_3	2100	>1000	Very good	Attacked slowly by dilute alkali	D-gun, plasma spraying, chemical vapour deposition (CVD)
Cr_2O_3	2400	>1000	Very good	Inert in most environments	D-gun, plasma spraying
Carbides					
eg TiC	3200	1000	Very good	Attacked by oxidising acids	Chemical and physical vapour deposition
VC	2600	500	Very good	Attacked by oxidising acids	Thermochemical
Nitrides					
eg TiN	2000	1000	Very good	Attacked by oxidising acids	Chemical and physical vapour deposition
Borides					
eg FeB	1650	200	Moderate	Attacked by mineral acids	Thermochemical treatment
CrB	3500–4000	800	Good	Attacked by HCl	Fusion of powder coatings

Figures in brackets are approximate hardnesses of dispersed carbide particles

(i) Hardness

As has been discussed in Chapter 1, a prime factor in the ability of materials to resist the various types of wear is high strength which in homogenous materials can be equated with high hardness:

$H = C\sigma_F$ where H is the indentation hardness
σ_F is the flow or yield stress
and C is a constant which is approximately three for hardened steels and about one for oxide abrasives.

Many of the materials used to resist wear, especially abrasive wear, however are not homogeneous and consist of hard particles, usually metal carbides, in a softer metallic matrix. In such materials abrasive wear resistance is a function of the hardness of both the dispersed phase and the matrix, the volume fraction of the dispersed phase and the size of the abrasive in relation to the size of the hard phase. More work needs to be done to understand fully these relationships.[5] The size of indentation produced on a macro hardness machine (eg Vickers) is many times the dispersion size which is usually in the range 5–100 μm so that an average hardness is obtained. The materials most widely used to resist wear are listed in approximate order of increasing macro-hardness in Table 2.1. The hardness required in a given situation is determined by the hardness of the abrasive or of the contacting surface, or by the magnitude of the imposed loads. In applications involving high loads the strength of the substrate is also important as otherwise the deformation in the substrate will exceed the fracture strain in the coating.

In situations involving adhesive wear the nature of the surface, especially its ability to resist local welding, is also important.

(ii) Impact resistance

The impact resistance of most wear resistant materials is low by normal engineering standards, but nevertheless differences in impact resistance between the vari-

Table 2.2 Salient features of surface treatment processes

Process	Approximate temperature of substrate during processing °C	Bond Strength, MN/m²	Porosity, %	Approximate rate of deposition or formation	Restriction on component shape
Thermal hardening	900	X	Nil		Axisymmetric best for induction hardening
Thermochemical treatments					
Carburising	925	X	Nil	0.1 mm/h	None
Nitriding	525	X	Nil	0.01 mm/h	None
Electrochemical treatments					
Chromium	50	350	Nil	0.03 mm/h	Not satisfactory on sharp edges
Co/Cr₃C₂ composite	50	350	Nil	0.03 mm/h	or blind holes
Chemical treatments					
Nickel phosphide deposition	90	350	Nil	0.015 mm/h	None
Phosphating	90	X		0.02 mm/h	None
Anodising	30	X		0.03 mm/h	None
Spraying processes					
D-gun	100	70–155*	1	0.3 kg/h	Accessible to a line of sight
Plasma	100	30–100*	1–3	2–5 kg/h	process
Wire gun	100	20	10–15	5–10 kg/h	
Spray and fuse	1050 (surface)	X	Nil	5–10 kg/h	
Welding processes					
Manual	1400 (surface)	X	Nil	3 kg/h	Accessible to line of sight process
Submerged arc	1400 (surface)	X	Nil	50 kg/h	
Chemical vapour deposition (CVD)	1000	X	Nil	0.001–0.01 mm/h	None
Physical vapour deposition (PVD)					
Evaporation	20	†		0.2–1 mm/h	Accessible to line of sight process
Sputtering	500	†	1	0.001–0.01 mm/h	Preferably accessible to line of
Ion plating	500	†	Nil	0.001–0.01 mm/h	sight process but some coverage of shadowed area

X Coating integral with substrate.

† Quantitative values not available. Evidence suggests that sputtered and ion plated coatings have bond strengths comparable to electroplate whilst evaporated coatings are much lower.

* Depends on material being sprayed and, in case of plasma, type of equipment used.

ous materials are significant in many applications. Although microstructure does have some effect on impact strength, in general the order of impact strength is the reverse of that of hardness. For example 12% Mn steel has much more impact resistance than 30% Cr cast iron, (100J and 2J respectively in a notched Charpy test) and, at the other extreme, aluminium oxide has much less resistance to impact than tungsten carbide-cobalt cermets (0.2 and 10J respectively in an unnotched Charpy test). In the great majority of wear situations impact forces of some magnitude are present and the selection of wear resistant materials is a compromise between hardness and toughness. In many cases tougher materials are preferable because failure by impact can be immediate and sometimes catastrophic, whereas inadequate hardness usually only results in a reduced life.

(iii) Strength and oxidation resistance at elevated temperatures

The maximum temperatures at which wear resistant materials can be used are listed in Table 2.1, Column 3. This temperature may be determined by softening of the material or by excessive oxidation. For example martensitic low alloy steels soften progressively above about 200°C and tungsten carbide oxidises rapidly when heated in air above about 550°C. Materials which can be used above 500°C include high chromium materials, eg 13% Cr steel, 30% Cr iron, chromium carbide-nickel-chromium cermets; cobalt alloys; nickel alloys and oxides.

(iv) Corrosion resistance

In Table 2.1, Column 4 a qualitative indication of corrosion resistance is given.[7] In general, materials with a high chromium, cobalt or nickel content have good corrosion resistance. Alumina and chromium oxide are also inert in most environments. However, if coatings are porous (eg sprayed coatings) or have microcracks (eg electroplated chromium), corrosion resistant undercoats or resin impregnation must be used to prevent corrodents reaching the substrate.

(v) Temperature of substrates

The approximate temperatures of the substrate during surface treatment are listed in Table 2.2, Column 2. High processing temperatures can lead to distortion so that with close tolerance components the surface layer must be thick enough to allow a post treatment machining operation. In high temperature deposition processes it is necessary to ensure that substrate and coating are metallurgically compatible so that any change in properties of either is kept within acceptable limits. For example it is not possible to apply weld overlays to titanium compressor blades to reduce fretting because of solution of titanium in the overlay and the effect of the high treatment temperature on the properties of the titanium.

Because of interdiffusion between coating and substrate high processing temperatures lead to high bond strengths.

(vi) Bond strength

Approximate values of the bond strength of overlay coatings deposited on a relatively cold substrate are shown in Table 2.2, Column 3. In process selection it is necessary to ensure that the bond strength of these coatings is adequate in applications involving concentrated subsurface stresses induced for example by misalignment, rolling or impact forces.

In welding processes, fusion of the overlay to the substrate takes place and the bond strength is equal to or greater than the strength of the substrate. Thermochemical treatments produce a hardened layer which is an integral part of the component.

In chemical vapour deposition, treatment temperatures are usually high (~1000°C) so that interdiffusion of overlay and substrate takes place resulting in very high bond strengths, again of similar magnitude to those of the substrate.

High bond strengths can in some applications be a disadvantage. For example if it is necessary to remove a coating prior to reclamation, this is much more difficult with a fused than with a sprayed coating.

(vii) Coating porosity

In processes involving spraying of powders the coating is porous the porosity ranging from less than 1% for most D-gun coatings, to 1-3% for plasma coatings and to 10-20% for combustion gun coatings. This porosity has several effects

(a) It weakens the coating structure so that coating hardness is lower than the hardness of the material in bulk form (plasma sprayed Al_2O_3 ~1000 HV cf bulk alumina ~2000 HV). However, in applications involving low loads, the abrasion resistance is frequently dependent on the basic material hardness rather than on the average coating hardness.

(b) Even at low values of porosity (~1%) penetration of the coating by gases and liquids may occur, so that such coatings cannot be used in corrosive environments unless sealed by organic resins or provided with a corrosion resistant underlayer.

(c) It can act as an oil reservoir, so preventing component failure in the event of temporary lubricant starvation.

(viii) Component shape

Some coating methods are unsuitable for use on complex shaped components. For example electrodeposits cannot be applied to components with sharp edges or blind cavities. Spraying processes are preferably carried out with the spray stream impinging at a high angle to the component surface. Most welding processes are also only applicable to relatively simply shaped components. Vacuum coating is essentially a line of slight process but in sputtering and, to a greater extent in ion plating, some coverage of areas not directly exposed to the source takes place.

In contrast, chemical vapour deposition and electroless nickel plating are noted for their ability to coat complex shaped components with a layer of uniform thickness.

(ix) Component size

In processes such as CVD and PVD the component size is limited by the size of chamber available. At present this is approximately 1 m³ but the size of components currently being treated is much smaller than this. In high temperature processes, especially those involving a subsequent heat treatment, component size is limited by the amount of distortion which can be tolerated. Nitriding carried out below the ferrite/austenite transformation temperature produces little distortion and gears up to two metres in diameter can be treated. With spraying and welding processes there is no upper limit to the size of component and there are variants of each process which can be used on-site. In electrodeposition component size is only limited by the size of bath available.

(x) Rates of coating deposition or layer formation

Rates of coating deposition or layer formation are listed in Table 2.2, Column 5. With chemical processes the rates are determined by the laws of chemical kinetics, and the size of plant available governs the area treated in unit time. With mechanical processes, such as spraying and welding, the amount of material deposited per hour is determined by the size and power of the equipment. In physical vapour deposition processes the rate of deposition is determined by the energy input to the coating materials reservoir or by its rate of extraction from the substrate during coating. As it is uneconomic to produce thick coatings at a low rate, there is a direct correlation between rates of layer formation and the thickness of the layers normally produced.

(xi) Coating thickness

In coating processes involving the deposition of hot

Fig. 2.2 Typical thicknesses of layers and coatings produced by surface treatments.

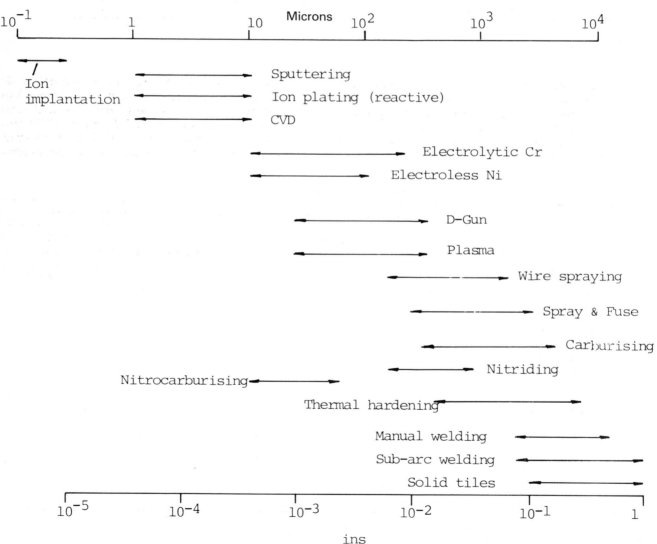

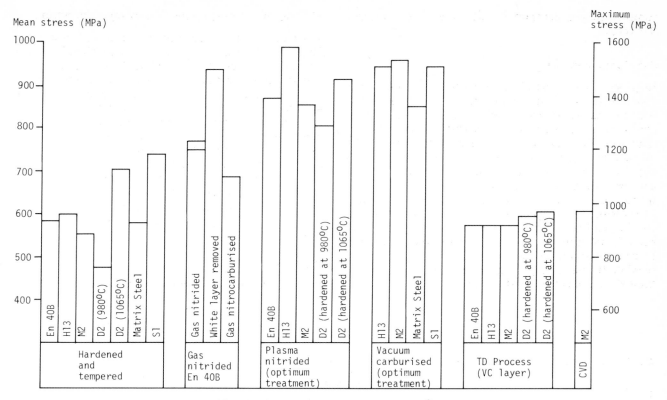

Fig. 2.3 Effect of surface treatments on unidirectional fatigue strength of various steels.[9]

material on a cold substrate, stresses are produced at the interface due to differential thermal contraction. In relatively hard dense coatings of oxides or carbides produced, for example, by D-gun and plasma deposition, these stresses can not be relieved by plastic deformation and coatings are limited, because of cracking, to thicknesses less than approximately 0.4 mm. Internal stresses due to dissolved impurities in electroplated chromium also impose a limitation on coating thickness to about 0.5 mm. In PVD processes hard ceramic coatings are limited in thickness to the 1–10 μm range by the slow rate of deposition.

Table 2.3 Effect of some surface treatments on the properties of a steel substrate[8]

Coating on steel C 45 W 3* Hardened and annealed	Adhesion resistance to			Abrasion resistance to			Fatigue strength		Corrosion resistance	
	itself	X 155 CrVMo 12 1*	Al$_2$O$_3$	flint	corundum	SiC	Bending	Rotary bending	SO$_2$-containing atmosphere DIN 50 018	Sodium chloride solution DIN 50 021
Without coating	–	–	+	–	–	–	o	o	– –	– –
Chromium		–	–	+	–	–	+	–	+	–
Nickel, chem. Age-hardened	+	+	+	–	–	–	+	+	+	+
Nickel, chem. Not age-hardened	–	–	+	–	–	–			+	+
Iron nitride	+ +	+	+	+	–	–	+ +	+ +	+	+
Iron boride	+	+	+	+	+	–	–	–	– –	– –
Titanium carbide (CVD)	+ +	o	–	+	+	+	o	o	+	+
Chromium carbide (CVD)	+ +	o	–	+	+	–	o	o	+	+
Tungsten carbide (CVD) Ni-intermediate layer		–	–	+	+	+	o	+	+	–

+ + very high   + high   o medium   – low   – – very low
*See Appendix 2 for steel compositions

These coating processes can obviously not be used economically in applications involving gouging abrasion where there is heavy wear, ie of the order of several millimetres, or in applications where high subsurface stresses occur. In such cases weld deposits and thermal or thermochemical treatments which produce thick hard layers would be considered. Alternatively, if only a small amount of wear can be tolerated as in, for example, a metal cutting tool or a precision drill spindle, then a thin coating is adequate.

The thicknesses of the layers and coatings which are normally produced by the various processes are illustrated in Fig. 2.2.

(xii) Effect of surface treatment on the fatigue strength of the substrate

A qualitative comparison of the effect of some surface treatments on the fatigue strength of a steel is given in Table 2.3.[8] This table also illustrates, *inter alia*, the effect of the hardness of the abrasives flint ($\sim$ 1000 HV) corundum ($\sim$ 2000 HV) and silicon carbide ($\sim$ 2500 HV) on the abrasive wear resistance of surfaces of varying hardness.

Thermal and thermochemical processes such as carburising and nitriding introduce compressive stresses to the surface layers which have a beneficial effect on fatigue strength. Some data on a variety of tool steels are given in Fig. 2.3.[9]

Electroplating, eg of chromium and nickel, produces residual tensile stresses at the surface and the effect is to reduce the fatigue strength of most steels. Results for chromium plate on a range of steels of varying strength is shown in Fig. 2.4.[10] The fatigue strength of chromium plated steels remains approximately constant ($\sim$ 20 tons/in^2) irrespective of the fatigue strength of the underlying steel.

A number of investigations have been made of the effect of sprayed coatings on fatigue strength.[11,12] Many of the results reported have been inconsistent. During spraying tensile stresses can be produced at the substrate surface, the magnitude of which depends on the expansion coefficient of the coating, coating thickness, the temperature of the substrate during

Fig. 2.4 Effect of electroplated chromium on the fatigue strength of steels.[10]

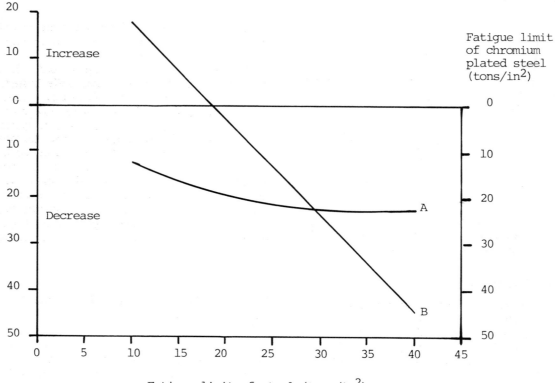

Effect of chromium plating on fatigue limit %

A Fatigue limit of chromium plated steels of varying strengths

B Change in fatigue limit after chromium plating of steel substrates of varying strengths.[10]
    (Thickness of chromium 0.006 in., internal tensile stress 4 tons/in^2).[10]

Table 2.4 Approximate costs of surface treatments

(a) Surface treatments

Treatment	Cost in £'s for coating a cylinder 50 mm dia × 500 mm long		Layer thickness	
	in Sweden[13] in 1979*	in UK in 1984	in Sweden	in UK
Carburising	–	2	–	0.75 mm
Nitriding	5.6	3	10 µm	0.1 mm
Nitrocarburising	2	4	10 µm	25 µm
Chromium (electrolytic)	3.2	8	50 µm	50 µm
Cobalt + Cr_3C_2 (electrolytic)	45	50	10 µm	0.3 mm
Nickel (electroless)	7.2	13	25 µm	50 µm
Plasma sprayed WC-Co	48	60	0.3 mm	0.3 mm
Al_2O_3		30	–	0.3 mm
Combustion gun spraying				
13% Cr wire	20	25	0.7 mm	0.7 mm
Ni-Cr-B spray and fuse	–	55	–	0.7 mm
Ni-Cr-B + (WC) spray and fuse	–	115	–	0.7 mm
Welding				
Iron base	32	76	4 mm	4 mm
Cobalt base	120	140	4 mm	4 mm
CVD TiC	80	80	10 µm	5 µm
PVD TiN	–	50	–	4 µm

* Converted 12.5 SKr = £1

(b) Solid tiles

Material	Thickness mm	Cost* £/m²
Alumina (92% Al_2O_3)	25	140
Alumina (97.5% Al_2O_3)	18	300
Fused Basalt	25	30
Ni-hard	25	120
13% Mn steel	12	70
Tungsten carbide-cobalt	3	1900

* Material cost only

spraying and rates of cooling. These tensile stresses together with defects in the coating can lead to a reduction in high cycle fatigue strength.

(xiii) Costs of surface treatments

Approximate costs of treating a steel cylinder 50 mm diameter and 500 mm long are given in Table 2.4. One set of figures was obtained in Sweden in 1979[13] and the other in the UK in 1984. Considerable variation in costs occurs between operators because of variations in the cost of process materials obtained from different suppliers and in labour costs. In batch processes, such as thermochemical and electrochemical treatments, costs vary considerably with batch numbers. For the UK costs in Table 2.4 it has been arbitrarily assumed that there were 100 components per batch.

The relationship between process costs and process selection is complex. Factors involved include the cost of the component, the ease with which it can be replaced, the cost of machine downtime, and the costs involved in redesigning to avoid or ameliorate the particular wear situation.

With components which are produced in large numbers, for example in the automobile industry it is usually more economic to design them so that acceptable performance is obtained using low cost surface treatments. On the other hand in aircraft gas turbines, performance, component integrity and increased times between overhauls permit higher process costs.

The effect of component cost is illustrated by cemented carbide tool tips coated by chemical vapour deposition. In this case the coating represents only a small additional premium on the relatively high basic tip cost. However, such expensive coatings would probably be unacceptable on cheaper components.

5. Summary of factors in surface treatment selection

The various factors influencing the selection of coatings and surface treatments to resist the wear processes described in Chapter 1 are summarised in Table 2.5. From a knowledge of the wear mechanisms operating in the various wear processes the essential surface requirements can be deduced and these are listed in Column 2 of the table. The main types of coatings and surface treatments meeting these basic requirements are given in Column 3. Restrictions on coating or treatment selection are imposed by both operational and environmental factors and the effects of some of these are listed in Columns 4 and 5 respect-

Table 2.5 Summary of factors involved in surface treatment selection

Wear process	Primary surface requirements	Principal candidate surface treatments	Specific operating conditions influencing treatment selection	General environmental conditions influencing treatment selection
Fretting	Metals of dissimilar composition to contacting surface. Non metallics. Hard materials	Sprayed Cu alloys, anodising (on. Al and Ti) nitrocarburising (on steels), plasma sprayed carbide cermets, electrodeposited cermets.	*Degree of lubrication* Use of solid lubricants (eg MoS_2) frequently obviates need for coatings *Amplitude of vibration* Cu alloys suitable at low amplitudes At high amplitudes hard coatings required.	
Contact fatigue	High yield strength, adequate toughness, thick coating, good bond strength.	Thermal and thermochemical treatments; weld deposits, sprayed and fused coatings.	*Applied stress* Thick case, usually obtained by carburising or thermal hardening but sometimes by nitriding, required at high operating stresses. At low stresses the cases obtained by nitrocarburising are adequate.	
Adhesive wear	Metals of dissimilar composition to contacting surface. Non metallics. Hard materials.	Sprayed Cu alloys: Mo. Thermochemical treatments; phosphating, plasma sprayed carbide cermets, electroless Ni. TiC (by CVD) TiN (by PVD)	*Degree of lubrication* Soft dissimilar coatings can be used where lubricant is present. Harder materials necessary when lubrication is marginal or absent. Hardness requirement determined by contact pressure and hardness of contacting surface. *Surface roughness* Wear greater with rougher surfaces. Non-metallic surfaces, produced by phosphating and nitrocarburising, prevent wear during 'running-in' at high loads and speeds.	Elevated temperatures (see Table 2.1) Thermal and most thermochemical treatments start to soften above 200°C. Above 500°C choice restricted to high Cr and Co alloy weld deposits, plasma sprayed Co alloys and Cr_3C_2-Ni-Cr cermets, electrodeposited Co/Cr_3C_2 sprayed and fused Ni alloys Corrosive environments (see Table 2.1) Sprayed coatings require complete sealing with organic resins or corrosion resistant bond coat to prevent penetration to substrate. Best corrosion resistance with Cr, Co or Ni alloys. Electroless Ni outstanding.
Percussive wear	High yield strength, adequate toughness, good bond strength.	Weld deposits, thermal and thermochemical treatments, D-gun carbide cermets, plasma sprayed carbide cermets and Co alloys. Electrodeposits (Co/Cr_3C_2)	*Impact stress* At low stresses materials of high hardness can be used. At higher stresses toughness becomes of increasing importance.	
Low stress abrasion	High hardness	All hard coatings	*Abrasive or abrading surface hardness* Coating selected must be harder than abrasive or contacting surface hardness but little advantage in increasing hardness above 1.3 times abrasive hardness.	
Machining wear	High hardness	All hard coatings	*Impact loads* Ancillary impact loads may restrict choice to tougher materials.	
Grinding wear	High hardness, adequate toughness, moderately thick coating.	Weld deposits. Thermal and thermochemical treatments.	Low stress abrasion and gouging conditions both applicable.	

Wear process	Primary surface requirements	Principal candidates for surface treatments	Specific operating conditions influencing treatment selection	General environmental conditions influencing treatment selection
Gouging	Adequate toughness, high hardness, thick coating	Weld deposits	*Size of abrasive particles* Tougher materials required for larger sized abrasives. Harder materials for smaller and/or harder abrasives.	
Erosion High angle impact	High hardness, adequate toughness, good bond strength	Weld deposits. Plasma sprayed carbide/cermets and metals.	*Velocity of impact* At low velocities (100 m/s) elastic materials (rubbers) can be used but at medium and high (250 m/s) velocities a combination of hardness and toughness is required.	
Low angle impact	High hardness	All hard coatings	Coatings must be harder than eroding particles at all velocities.	

ively. However, there is a wide range of conditions under which wear can occur, a large number of surface treatments available and a limited amount of information on the performance of these treatments under controlled conditions. An analytical approach is therefore not always possible and processes are selected on the basis that they have proved successful in analogous situations.

Some further comments on process selection as applied to hardfacing by welding are made in Chapter 11.

References

[1] THELNING, K E, *Steel and its heat treatment*. Butterworths, London, 1984.

[2] FICHTL, W, Boronising and its practical applications. *Materials in Engineering*, Vol. 2, p 276, December 1981.

[3] ARAI, T and FURAMA, T, Carbide surface treatment of dies. *Paper No G-T81-092. Society of Die Casting Engineers*, 1981.

[4] DEARNALEY, G, The ion implantation of metals and engineering materials. *Trans Inst Met Finishing*, Vol 56, 1, p 25, 1978.

[5] MOORE, M A, Abrasive wear by soil. *Tribology International*, Vol 8, 3, 1975.

[6] Corrosion, ed Shreir, L L. Newnes, London, 1965.

[7] EVANS, U R, *Corrosion and oxidation of metals*. Arnold, London, 1977.

[8] HABIG, K H, Wear, corrosion and fatigue behaviour of steels coated with hard surface layers. *Conf. on Wear of Materials*, Reston, Va, p 288, 1983.

[9] CHILD, H C, PLUMB, S A and REEVES, G. The effect of thermochemical treatments on the mechanical properties of tool steel. *Heat treatment of Metals*, Vol 9, 4, p 93, 1982.

[10] WILLIAMS, C and HAMMOND, R A F. The change of fatigue limit on chromium or nickel plating. *Trans. Inst. Met. Finishing*, 34, p 317, 1957.

[11] KESHTVARZI, A and REITER, H. The effect of flame sprayed coatings on the fatigue behaviour of high strength steels. *10th International Thermal Spraying Conference*, p 229, 1983.

[12] BERTRAM, W and SCHEMMER, M. Fatigue behaviour of thermally coated steels. *10th International Thermal Spraying Conference*, p 242, 1983.

[13] CHILD, H C. Report on 2nd International Congress on Heat Treatment. *Heat Treatment of Metals*, Vol 9, (4), p 91, 1982.

CHAPTER 3

Applications

A number of components have been selected from a range of industries to illustrate the use of surface treatments in resisting wear. These together with the treatment(s) used in each case are listed in Table 3.1. As far as possible components have been selected where more than one treatment has been used and information is available on the effect of the treatment on performance.

Further applications of the various treatments are discussed in Chapters 4 to 11.

1. Gears

The load carrying capacity of a gear material is determined by its ability to sustain contact stresses on the tooth flanks and bending stresses at the base of the teeth. Besides requiring resistance to surface fatigue by compressive stresses, gear materials require resistance to Hertzian shear stresses below the surface, which reach a maximum value at a depth which increases with load and increasing tooth flank radius. A thick layer is thus required and, as Hertzian stresses can reach values of 1500 MN/m^2, virtually all treatments are eliminated from consideration except thermal and thermo-chemical, where the surface is an integral part of the component.

The bending fatigue strength of notched specimens of through hardened steels increases with increasing material tensile strength, up to 900–1000 MN/m^2, above which there is a reduction in fatigue strength due to an increase in notch sensitivity. Two types of stress parameter are used in the design of gears; S_{ac} the allowable contact stress and S_{at} the allowable bending stress at the tooth root. Both parameters provide stress levels giving expected lives of 10^7 cycles or greater. Typical design data for various gear materials and treatments are given in Table 3.2. Gears also require resistance to wear and scuffing; wear behaviour is mainly determined by surface hardness, surface roughness, material compatibility and lubrication. The subject of gear wear is complex and has previously been discussed at length, the state of lubrication being a major factor.[4]

Thermal hardening of tooth surfaces can be carried out by induction or flame heating with appropriate quenching. Induction hardening can be of two forms. Contour hardening, which produces a uniformly deep hardened case, is a method used for medium to large gears, in which a tooth-shaped inductor traverses the tooth spaces which are submerged in quenchant. With careful process control favourable residual stresses are obtained in the case. This process has been particularly successful for large gears employed in the mining industry, eg ball-mill drive pinions. In the alternative induction hardening method and in spin-

Table 3.1 Some applications of surface treatments for wear resistance

Application	Thermal	Thermochemical	Electrochemical	Chemical	CVD	PVD	Spraying	Welding	Solid Tiles
Gears	✓	✓							
Steering gear	✓								
Cylinder liners	✓		✓						
Piston rings		✓	✓				✓		
Poppet valves	✓		✓					✓	
Aero engine components			✓				✓	✓	
Textile machine components			✓	✓			✓		
Metal cutting tools (carbide)					✓				
Metal cutting tools (HSS)					✓	✓			
Metal forming tools		✓		✓	✓	✓	✓		
Rock, ore and earth engaging equipment	✓							✓	
Chute liners (steel plant)								✓	✓
Pulverised fuel handling equipment							✓	✓	✓

flame hardening, the gear is rotated inside a circular inductor or a ring of gas jets, and the teeth are then through hardened by quenching. The bending fatigue strength can be reduced by this type of treatment, due to the increased notch sensitivity of the through hardened teeth, but by allowing a heat diffusion period, at a reduced heating level, the root areas may be hardened to give favourable residual stresses with good fatigue strength.

Thermochemical treatments are widely used for hardening gears. Carburising of carbon steels and low alloy steels offers a wide range of easily controlled case depths of hardness 600–800 HV, with the devel-opment of beneficial compressive residual stresses. Carburised gears are used in high duty automotive, industrial, marine and aircraft applications. A disadvantage with carburising is the growth and distortion that occurs, which is particularly serious with large gears (eg above 1 metre diameter). Diametral growth of up to 0.3% may occur and be accompanied by out-of-roundness, out-of-flatness, tapers and helix angle changes. Distortion can be reduced by quenching in a press and/or corrected by grinding, but the latter is expensive and leads to non-uniformity of the final case depth. Grinding is always required to produce the accuracy and surface finish needed for high precision

Table 3.2 Design data for gear materials (Compiled from references 1, 2, 3 and 4)

Condition or treatment	Material	Surface hardness	Tensile strength of core	Allowable contact fatigue stress (S_{ac})	Allowable bending fatigue stress (S_{at})
		HV	MN/m^2	MN/m^2	MN/m^2*
Normalised	0.4%C steel	165	530	10	145
Through hardened	C-Mn, Mn-Mo, 3% Ni	200	695	12.5	159
	1% Cr-Mo, 3% Ni-Cr, 3% Cr-Mo	250	850	21.5	214
	1% Cr-Mo, 2.5% Ni-Cr-Mo	270	925	23.5	221
	3% Ni-Cr-Mo	365	1230	26	234
Carburised	C case hardening steels	800	495	65.5	214
	2% Ni-Mo	725	750	69	255
	3% Ni-Cr-Mo	750	950	75.5	283
	4.25% Ni-Cr, 4.25% Ni-Cr-Mo	710	1250	89.5	345
Nitrided	3% Cr-Mo	850	800	55	179
	3% Cr-Mo-V	850	1250	69	283
Induction or flame through hardened teeth	0.4%C steel	500	550	54	110
	1½% Ni-Cr-Mo	550	850	45	145
	2½% Ni-Cr-Mo	550	925	45	172
Induction contour hardening	0.4%C steel	500	530	45	166
	1½% Ni-Cr-Mo	550		45	207
	2½% Ni-Cr-Mo	600		45	255

*These values do not include stress concentration factors which may be appropriate at the tooth roots.

Table 3.3 Comparative gear scuffing resistance of steel with various thermochemical and thermal treatments (IAE gear machine)[5]

Materials for gear pair		Average scuffing load, N/mm of face width	
		SAE 50/60 oil (undoped)	SAE 10 oil (doped)
Nitrided 3% Cr-Mo-V	Nitrided 3% Cr-Mo-V	1550	1270
Carburised 2% Ni-Mo	Nitrided 3% Cr-Mo-V		1900
	Carburised 2% Ni-Mo	800	940
	Air hardened 4¼% Ni-Cr		520
Carburised 3% Ni-Cr	Carburised 3% Ni-Cr	820	760
	Oil hardened 1½% Ni-Cr-Mo		480
Carburised 4¼% Ni-Cr	Nitrided 3% Cr-Mo-V		1050
	Carburised 4¼% Ni-Cr	680	720
	Air hardened 4¼% Ni-Cr		570
Oil hardened 1½% Ni-Cr-Mo	Oil hardened 1½% Ni-Cr-Mo	800	580
Oil hardened 4¼% Ni-Cr	Nitrided 3% Cr-Mo-V		1160
	Air hardened 4¼% Ni-Cr	610	455

Fig. 3.1 Internal ring and spur gears treated by plasma nitriding.

gears. Profile grinding must be controlled to avoid; (a) overheating with impairment of the residual stress pattern; and (b) stress-raising steps near the tooth roots.

Nitriding of steel produces a thin case (up to ~0.7 mm) with favourable residual stresses and, since the steel is not heated to the transformation temperature or quenched, distortion is low. Nitrided gears may therefore be preferred for high duty applications even though the fatigue strength may be lower than for carburised gears (due to slightly less favourable residual stresses). For large nitrided gears, where the maximum Hertzian shear stress is near or below the case/core interface, high core strength steels are used, hardening and tempering being carried out prior to nitriding. Nitriding produces an undesirable brittle 'white-layer' of iron nitride at the surface but with correct material selection and processing conditions it can be minimised. Although the white layer is sometimes removed, most nitrided gears are put into service without further machining or chemical treatment. Nitrided gears have excellent scuffing resistance (see Table 3.3).

Plasma (or ion) nitriding is an alternative to gas nitriding which has been applied to large high precision gears. For example the gears (~1 m diameter) used in the power transmission systems of coal seam cutting equipment (Fig. 3.1) were initially manufactured from En8 steel which was flame or induction hardened to confer the required abrasion and fatigue resistance.[6] However, these techniques result in considerable distortion and require the use of expensive grinding equipment to restore the gears to original dimensions. At low production levels the investment in this grinding equipment proves to be uneconomic. Plasma nitriding has produced the required surface properties and allowed distortion to be controlled to

a level at which machining is not required. Steels such as En 40B, En 19 and En 24 are treated for up to forty hours in cracked ammonia at 530°C at which temperature the high core strength produced by hardening and tempering is maintained. Surface hardnesses of 850 HV are obtained with a total case depth between 0.5 and 0.6 mm and an iron nitride layer 8 μm thick. Distortion across a 150 mm tooth flank is less than 25 μm.

Carbonitriding is a short duration process giving a relatively shallow hardened case which confers improved contact fatigue strength and wear and scuffing resistance to carbon or low alloy steels. It is therefore an attractive process for the cheaper, smaller size gears. Nitrocarburising processes (eg Tufftride, Sulphinuz) are useful low temperature (570°C) treatments which confer good wear and scuffing resistance, provide some improvement in contact fatigue strength, do not require special steels and introduce minimal distortion problems. For example, for SAE4140 steel, the Tufftride process typically gives increases in the contact fatigue strength of 100% and increases in the bending fatigue strength of 30%.

Low temperature chemical treatments, such as phosphating and Sulf BT, are used principally to improve the wear and scuffing resistance of gears during their running-in period. These processes may be applied to carburised or heat treated gears without risk of tempering. To minimise adhesive wear and scuffing it is normal practice to make gear pairs of differing surface hardness, usually by making the pinion harder than the wheel.

The risk of gear scuffing is highest during running-in, eg with gears in the 'as-machined' condition. Wydler[7] has shown (Fig. 3.2) that the load to cause scuffing of test gears is increased following extended running at a safe load. Although gears with a rougher

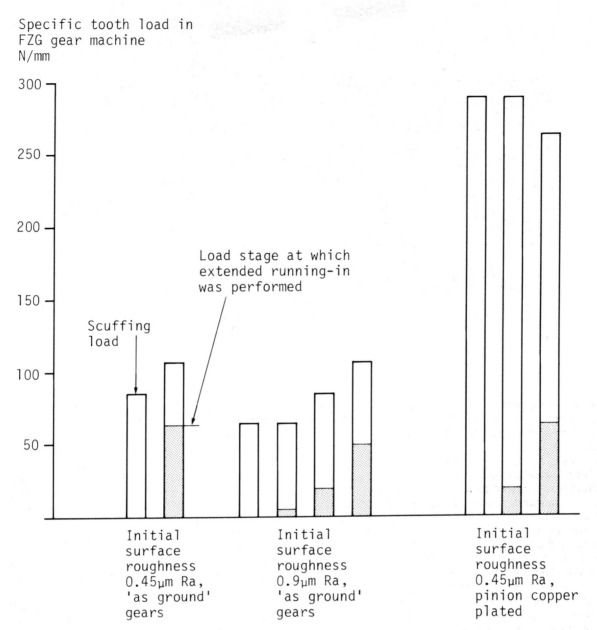

Specific tooth load in
FZG gear machine
N/mm

Fig. 3.2   Influence of surface roughness, running in and electrolytic copper plating on the scuffing resistance of carburised gears in the FZG test.[7]

machined surface had a lower scuffing load, their performance following running-in was the same as that of smoother gears. Soft metal platings, eg copper and silver, up to 0.01 mm thick are often applied to pinions to increase the scuffing resistance of 'as-ground' gears. Copper plated FZG test gears (Fig. 3.2) had scuffing loads about four times those of unplated gears, no further improvement being obtained on running-in. Plating can however adversely affect fatigue strength and tooth breakages have been attributed to this effect.

2. Automobile steering gear

Most applications of laser thermal hardening have so far been on automobile components since high volume production and minimum distortion requirements enable the technique to be economically viable. For example, since 1974 Saginaw Steering Gear Division of General Motors has been using CO_2 lasers to internally harden steering gear housings.[8] The housing unit had been originally designed to operate at a stress of 4.8 MN/m^2 but subsequent design changes had increased this to 10 MN/m^2. This increased loading created a wear problem. The housing units are produced in ferritic malleable cast iron, and since machinability, impact properties, and tensile fatigue strength are adequate to meet manufacturing and performance requirements, designers were reluctant to use alternative materials (such as pearlitic cast iron) which, although giving higher cast hardness, were more expensive and lacked the desired machinability.

Various conventional treatments including direct hardening, Tufftriding, nitriding and induction hardening were investigated to promote greater wear

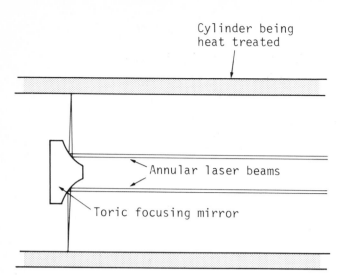

Cylinder being heat treated

Annular laser beams

Toric focusing mirror

Fig. 3.3 Laser hardening of internal bores.

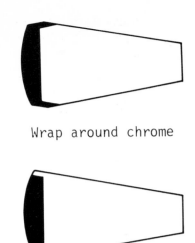

Wrap around chrome

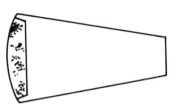

In-groove chrome

In-groove plasma

Fig. 3.5 Piston ring design profiles with various coatings.[13]

resistance but proved unsatisfactory because of distortion. The problem was overcome by using a laser to produce several wear tracks along the cylinder bore at points of greatest mechanical loading. Fig. 3.3 shows how laser beams can be deflected by mirrors to selectively heat the internal bores of cylinders and Fig. 3.4 is a photograph of a section of the housing showing the hardened tracks. The tracks measure 1.5–2.5 mm wide by 0.25–0.35 mm deep and exhibit a hardness of 60 HRC. Since less than 10 g of the 6.3 kg housing is hardened, there is no distortion. This new concept of wear patterns rather than overall hardening has proved successful, field trials having shown a ten-fold increase in wear resistance.

3. Cylinder liners and piston rings

The functions of piston rings are to provide a sliding gas seal with the cylinder wall and to prevent excessive amounts of oil from entering the combustion chamber. In normal operation the ring/cylinder couple is lubricated hydrodynamically and is only subject to mild abrasive wear. However, if lubrication is marginal, during periods of high acceleration or overload, then adhesive wear (scuffing) can occur. Conditions are more severe in diesel than in petrol engines because

Fig. 3.4 Cut away section of steering gear housing showing laser hardened strips.

of the higher pressures and temperatures and the demand for both longer lives between overhauls and lower oil consumption. Diesel fuel has also, in general, a higher sulphur content so that corrosion may increase the amount of abrasive wear.[9] Piston rings operate under more severe conditions than the cylinder liner in that they are subjected to continuous wear under varying load whereas any one part of the liner is only subjected to intermittent wear.

(i) Cylinder liners

In early engines grey phosphoric irons were used for both liner and rings, a requirement being low ferrite content (<5%) to minimise the incidence of scuffing. Most cylinder liners are still cast iron, some increases in strength and wear resistance having been obtained by small additions of carbide stabilising elements such as Cr, Ti, V and Mo. Wear resistance can be increased by induction hardening the bore to 400–500 HV. Chromium plating was formerly used on diesel engine cylinder bores mainly to reduce corrosion from high sulphur fuels. However, the development of lubricants with sufficient alkali to prevent corrosion of cast iron liners when burning fuels with sulphur contents as high as 5% has removed the need for chromium

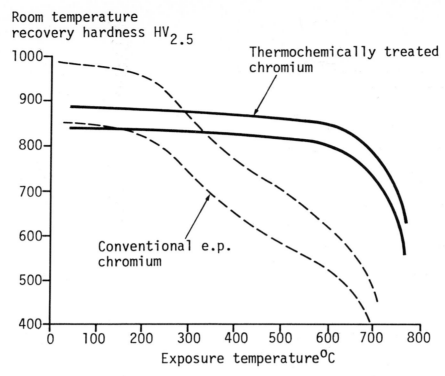

Fig. 3.6   Comparative thermal softening of conventional electroplated chromium and after thermochemical treatment.[14]

plating. Chromium plating can however be used to restore worn cylinder liners to original size and this is cost effective in some two stroke locomotive diesels.[10] It is also used on steel cylinder liners which are used on some high performance engines. To improve the oil retention properties of the chromium plate it may be etched electrolytically (van der Horst process) to form a series of interconnected channels or simply shot blasted.

(ii) Chromium plated piston rings

As chromium plate has a low resistance to adhesive wear against itself the use of chromium plated cylinder liners prevents the use of chromium plate on piston rings, the component subjected to the greatest wear. The more normal arrangement in an internal combustion engine is therefore a cast iron cylinder liner and a chromium plated cast iron compression ring. The other rings which are subjected to less severe wear conditions may also be chromium plated.[9, 11] The properties of chromium plate which make it attractive are its high hardness (~ 1000 HV) to give good abrasion resistance, its low friction values against cast iron, and its corrosion resistance.

Coatings on piston rings range in thickness from 0.1 mm for small engines to 0.3 mm for large engines or off highway vehicles operating in highly abrasive environments. The coatings are treated by one of the methods mentioned above to improve oil retention properties.[12] Depending on ring design they may be deposited as wrap around coatings or as inlays[13] (Fig. 3.5).

Chromium plated piston rings are used on the vast majority of engines but, where lubrication is marginal

or temperatures are high, eg in highly rated turbocharged diesels, failures have occurred by scuffing. Several alternative treatments are therefore being explored. For example it has been shown that if chromium plate is nitrocarburised in a salt bath a network of nitride is formed throughout its thickness which prevents the high temperature softening which leads to adhesive wear[14] (Fig. 3.6). A proprietary lapping technique known as plateau honing and which leaves a pitted surface is also claimed to increase scuff resistance.[15]

(iii) Flame sprayed piston rings

Spraying is the method which is being most actively explored for the development of ring coatings with a high resistance to scuffing (Table 3.4). Flame sprayed molybdenum, probably because it maintains its hardness at elevated temperatures, has notable anti-scuff properties and was introduced as a piston ring coating in the 1960s. The coating, which is laminar in structure and has a porosity of about 15%, has in general been highly successful but delamination of the coating sometimes occurs. To improve coating strength a material was developed which, when plasma sprayed, consisted of molybdenum particles in a matrix of a dense nickel chromium silicon boron alloy. This coating is used successfully in engines with ratings up to 150 BMEP. (Brake mean effective pressure.)

Several other plasma sprayed alloys have been studied. For example some of the molybdenum in the above alloy has been replaced by chromium and iron to produce a cheaper coating, and chromium carbide particles have been added to molybdenum to improve wear resistance.[13] Some of the properties of chrome

27

carbide-molybdenum alloys in comparison with chromium plate are shown in Figs. 3.7 and 3.8.

In addition to molybdenum based alloys, plasma sprayed ceramic materials (mainly 60% Al_2O_3–40% TiO_2) have been investigated. These have shown exceptional scuff and abrasion resistance but some have failed prematurely at the ceramic metal bond interface. Such coatings are expensive because diamond lapping is necessary to produce the required surface finish.

All plasma coatings are deposited in inlays (Fig. 3.5) so that there are no exposed edges but there is always the danger of scuffing of the cast iron bands. The performance of plasma sprayed coatings is much more sensitive to variations in processing than chromium plate. However, with plasma spraying the coating composition can be altered to suit specific applications.

One American group's view[13] of the relative merits of chromium plate and plasma sprayed coatings is given in Table 3.5.

4. Poppet valves in internal combustion engines

These valves are subjected to wear in three areas; the end of the stem or tip is worn by impact with the tappet, the stem by sliding wear between it and the bearing surface of the valve stem guide, while the sealing face at the valve head is subjected to erosion and corrosion by the hot combustion gases and by solid and liquid contaminants in them.[16] The severity of wear at these points varies with the duty, particularly at the head where the average temperature in the hottest part of the valve increases from about 600°C in a light duty engine, eg in a lawn mower, to about 900°C in a heavy duty truck. These temperatures occur at the centre of the valve rather than the sealing face which being cooled by frequent contact with the cooler valve seat is usually less than 500°C.

Tip hardness can be catered for by correct choice of material for the body of the valve, the recommended minimum tip hardness of 35 HRC (340–360 HV) being obtained by local induction hardening. Wear

Table 3.4   Development of piston ring materials for heavy duty diesel engines[12]

Material	Advantages	Limitations
Grey phosphoric iron } Malleable iron	Cheap. Can be used against liners of similar composition.	Distorts and scuffs at elevated temperatures.
Chromium plated (0.1 mm) iron	Low friction against cast iron liner. Long life when lubricated. Relatively cheap.	Scuffs under marginal lubrication conditions. Temperature limited. Corrodes in high sulphur fuel.
Wire sprayed molybdenum	Does not scuff even under extreme conditions.	More expensive than chrome plate. Coating can delaminate in heavy duty engines. Properties sensitive to application method.
Plasma sprayed molybdenum alloy (Mo + Ni-Cr-Si-B)	Higher strength than flame sprayed molydenum. Good scuff resistance.	More expensive than molybdenum.
Plasma sprayed molybdenum alloy with chrome carbide additions	Lower wear rate than straight molybdenum alloy.	Possibility of liner wear

Table 3.5   General comparison of chromium plated vs plasma coated top rings*

Parameter	Chromium	Plasma
Basic constituents	Electroplated hard or porous chromium	Infinite metallic or cermet powder combinations
Ring wear resistance	Excellent	Excellent
Cylinder wear resistance		
hard and soft iron	Excellent	Excellent
hard steel	Poor	Excellent
chromium plated	Cannot be used	Excellent
Scuff resistance	Good	Excellent
Adhesive bond	Excellent	Very good
Heat resistance	Poor	Excellent
Oil carrying capability	Poor	Excellent
High BMEP engines	Fair	Excellent
Fuel and lubricant	Restricted	Highly adaptable
Ring cost		
'wrap-around' design	Low	High
'in-groove' design	Moderate	Moderate

*References to plasma coatings apply to the most suitable plasma coating for the specific application.

Table 3.6 Relevant high temperature properties of some exhaust valve materials

Material	Hot tensile strength MN/m^2					Hot hardness (HV)				
	20°C	200°C	400°C	600°C	800°C	20°C	200°C	400°C	600°C	800°C
En 52	1003	927	818	340	62	300	280	265	220	75
21–4N	1050	957	834	618	371	320	270	230	200	130
Nimonic 80A	1220	1174	1158	1081	633	290	280	275	265	220
Nimonic 81	1050	973	896	865	587					

of the stem is reduced by hard chromium plating, thicknesses varying from 1–2 μm in petrol engines to 300 μm in heavy duty diesel engines. Hard chromium plate satisfies the requirement that the coating should have low friction and low wear rate with respect to the softer valve guide material, usually cast iron. The mechanical property requirements at the shoulders and in the head are adequate hot fatigue and creep strength and hot hardness. Wear resistance is required at the seal faces. The corrosive conditions here, and at the sealing face of the valve, require good hot corrosion resistance to, eg, lead compounds, sulphur dioxide, sulphates and vanadates, depending on the

Table 3.7 Material combinations in two part valves

Head	Stem
21–12 N	En 19C
21–4 N	En 18, En 24
Nimonic 80A	En 52, $\times$45 Cr Si 9.3
Nimonic 81	En 52

type and cleanliness of the fuel. Some of the valve materials in common use are, En 52, En 59, 21–12 N, 21–4 N, Nimonic 80A, Nimonic 81. The variation in tensile strength and hardness with temperature of

Fig. 3.7 Bar graph showing abrasive engine test results using BP additive.[13]

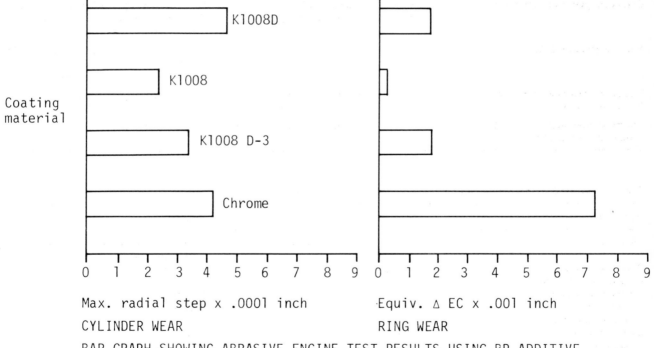

Max. radial step x .0001 inch
CYLINDER WEAR

Equiv. Δ EC x .001 inch
RING WEAR

BAR GRAPH SHOWING ABRASIVE ENGINE TEST RESULTS USING BP ADDITIVE

K1008—chrome carbide-molybdenum plasma sprayed coatings D-
   unspecified modification.
BP additive is silicone which on combustion forms silica which accelerates wear.
Δ E.C.—Average ring wear $\times 2\pi$

Table 3.8 Some typical alloys used for hardfacing exhaust valve sealing faces

Material	Nominal composition %								Hot hardness (HV)	
	Cr	W	C	Ni	Fe	Si	B	Co	600°C	750°C
Stellite 6	33	5	1	3 max	3 max	–	–	Bal	270	185
1	30	2.5	12.5	3 max	3 max	1	–	Bal	425	230
12	29	9	1.8	3 max	3 max	–	–	Bal	300	245
20	33	18	2.5	–	–	–	–	Bal	418	300
Deloro 60	16	–	0.5	Bal	4.5	4.5	3.5	–	340	115

some alloys are given in Table 3.6. Nimonics must be used in the hotter engines. The problems caused by having different types of wear at either end of the valve are being solved by use of two part valves, joined by friction welding, the head being of different composition to the stem. Some typical combinations are shown in Table 3.7.

Catering for the stresses in the neck of the valve will not necessarily solve the problems of wear at the sealing faces although the Nimonics with their higher chromium and nickel content and higher strengths are better able to resist mechanical damage and wear than the iron base alloys. Improvements in resistance to wear can be obtained by hardfacing locally at the sealing face. Hardfacing materials should have good corrosion resistance, good hot strength, be able to resist thermal fatigue, have a coefficient of thermal expansion near that of the substrate and be able to be deposited easily with minimum post-deposition machining or grinding. Table 3.8 shows some typical cobalt- and nickel-base alloys in common use. Hardfacing is usually applied to diesel engine exhaust valves. With large valves (> 50 mm diameter) cracking

of hardfacings can occur in service and is usually due to residual stresses introduced during deposition, which can be reduced by post weld peening. Other causes of cracking can be incompatible expansion coefficients between coating and substrate or thermal fatigue.

Stellite 6 or its equivalent is the most widely used hardfacing in this application but there is a trend to use harder materials, ie Stellites 12, 1 and 20. The higher hot hardness of these alloys resists indentation of the seal faces by hard particles carried over in the hot combustion gases. Oxyacetylene and TIG welding processes are used for hardfacing although both are being superseded by the plasma transferred arc process, which gives better quality coatings with controlled dilution and increased flexibility in choice of deposit material.

The large demand for poppet valves justifies the use of fully automated production lines with output of about 4000/h. Under these conditions high rates of deposition are needed with a high degree of control of powder feed and heat input. Laser hardfacing could satisfy this latter requirement.

Fig. 3.8 Bar graph showing 2 cycle engine test using high sulphur fuel.[13]

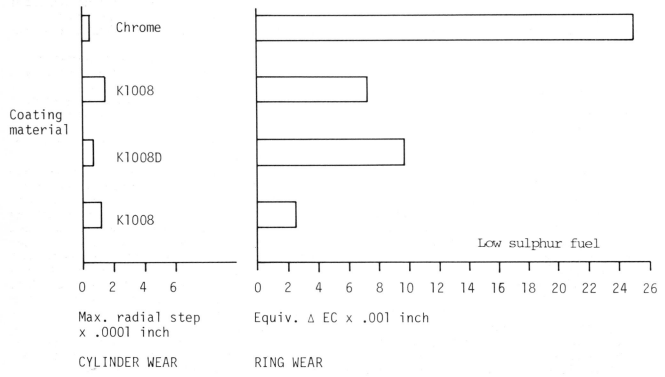

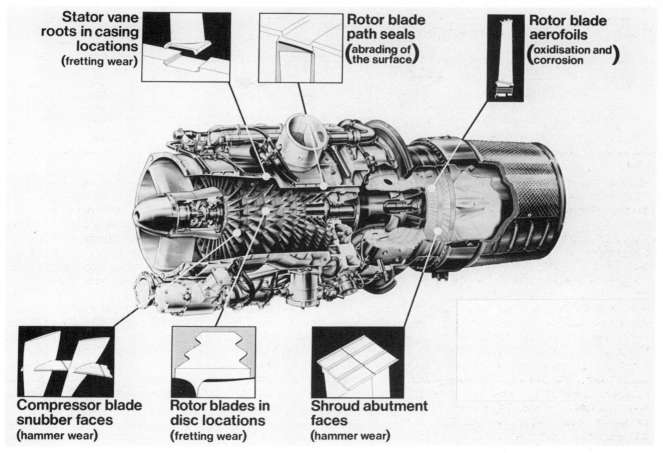

Stator vane roots in casing locations (fretting wear)

Rotor blade path seals (abrading of the surface)

Rotor blade aerofoils (oxidisation and corrosion)

Compressor blade snubber faces (hammer wear)

Rotor blades in disc locations (fretting wear)

Shroud abutment faces (hammer wear)

Fig. 3.9   Aero engine showing some wear situations.

Current practice is to coat iron base martensitic and austenitic valve alloys with cobalt base hardfacing alloys and to coat Nimonic alloys with nickel base alloys (eg Deloro 60). With the current trend to use more impure residual bunker fuels in ships' diesel engines there is a need to develop other less costly corrosion resistant hardfacing materials.

5. The aero gas-turbine

The main types of wear which occur in an aero engine either singly or in combination are:

> hammer wear
> fretting
> and sliding wear

These can occur over the temperature range from sub-zero at the air inlet to over 1000°C in the turbine section.[17, 19] Some typical wear situations are illustrated in Fig. 3.9. One of the main considerations in the selection of treatments to resist wear is that they have minimum effect on the mechanical properties of the substrate which may be nickel alloy, titanium alloy or steel. Another important factor is the limitation of possible damage to complex expensive components during processing and refurbishment.

Of the various treatments available spraying in its various forms comes closest to satisfying these demands and because of the wide range of materials which can be sprayed this process is used on over 600 different components in a typical aero engine. Of these approximately 50% are treated by flame spraying, 45% by plasma spraying and 5% by the detonation gun. Detonation gun coatings are applied in preference to plasma sprayed coatings where:

(a) higher bond strengths are required to resist high levels of hammer wear.

(b) spraying cannot be carried out normal to the surface, (under this condition D-gun coatings give acceptable levels of bond strength).

(c) the smoother surfaces obtained allow grinding to be dispensed with or replaced by a simpler brushing operation.

The coatings applied to some aero engine components are listed in Table 3.9.

In the aero engine the two main areas where hammer wear occurs are the snubbers on fan blades and the shrouds on turbine blades. In both areas some sliding wear also occurs. Coatings to resist hammer wear must, in addition to having high hardness, have some impact strength and high adherence to the substrate. Metal or cermet coatings are therefore used. On fan blade snubbers WC-Co cermets with Co contents in the range 10–15% applied by D-gun have been found to be effective. Initially harder coatings with 9% Co were used but these were found to fail by surface fatigue.[20] On turbine blade shrouds coatings such as Co-Cr-W alloys (Stellite type) and Cr_3C_2-Ni-Cr, which have better high temperature oxidation resistance than WC-Co, are used.

31

Table 3.9 Coatings applied to aircraft gas turbine parts

Component	Material	Coatings	Process	To provide protection against
Turbine blades: shroud surfaces	Superalloy	Cr_3C_2-NiCr Co-Cr_3C_2 composites Stellite alloys*	Plasma spraying or D-gun Electrodeposition Welding	Sliding and impact wear. Hot corrosion.
Turbine blades: blade roots	Superalloy	Co-Cr_3C_2 composites	Electrodeposition	Fretting
Turbine blades: aerofoil sections	Superalloy	Nickel aluminide/MCrAlY	Diffusion coating Vacuum coating Ion plating Plasma spraying	Hot corrosion
Compressor blades: final stage	Titanium	WC-Co	D-gun	Erosion
Compressor shaft: hub journal area	Titanium	WC-Co	D-gun or plasma spraying	Sliding wear
Fan blades: mid span stiffeners	Titanium	WC-Co	D-gun	Sliding wear and impact
Seals: main shaft	Carbon	Al_2O_3	Plasma spraying	Machining wear
Seals: face	Steel	Cr_3C_2-NiCr	D-gun or plasma spraying	Machining wear
Combustion chambers: locating/clamping rings, dowel pins	Superalloy	Cr_3C_2-NiCr	D-gun or plasma spraying	High temperature fretting

*on older engines

Fretting occurs between two surfaces which have nominally no relative movement between them but which vibrate over very small amplitudes. It can occur between compressor and turbine rotor blade roots and the corresponding discs. Where no machining of the coating is possible, because of the geometry of the component, soft coatings such as Cu-Ni-In and electroplated Co-Cr_3C_2 are used at low and at high temperatures respectively. Where machining is possible then a range of hard coatings, such as WC-Co and Cr_3C_2-Ni-Cr, applied by plasma spraying are used. In sliding wear which occurs at the interface between compressor stator blades and their casings the same hard coatings to reduce fretting wear are effective.

The properties of some of the coatings used in aero engines as determined in rubbing wear and hammer wear tests are illustrated in Fig. 3.10.[21]

In aero engines a range of abradable coatings to provide gas path seals are also used. A polyester resin with an aluminium filler which is plasma sprayed is used up to 330°C in LP, IP and part of the HP compressors. In the higher temperature part of the HP compressor a nickel-graphite abradable is used.

Coatings are also applied to turbine blades to resist oxidation. Pack-aluminising is still widely used. The newer MCrAlY type coatings which resist higher temperatures are applied by vacuum evaporation, or by shrouded or vacuum plasma spraying.

6. Synthetic fibre processing machines

Synthetic fibres such as nylon and polyester subject the parts of processing machinery, with which they come in contact, to low stress abrasion. The fibres are frequently lubricated with proprietary mixtures, some of which are corrosive. In earlier machines components were chromium plated over an undercoat of nickel for corrosion protection. Difficulties in obtaining uniform coatings on complex shapes led to the introduction of electroless nickel which could be heat-treated to a similar hardness to chromium (~1000 HV). With higher yarn speeds electroless nickel was found to have insufficient wear resistance and solid alumina ceramics were introduced for simple shaped components and plasma sprayed ceramics for those of more complex shape with considerable increases in component lives. In addition to being abrasion resistant, surfaces must also have controlled levels of yarn/surface friction.[22] Two types of surfaces are required, those with a low friction value for guidance and sliding contacts and those with high friction for traction. The former type is usually obtained by abrading the surface with brushes loaded with abrasive which gives a comparatively rough surface 2–3 μm CLA. High friction surfaces are obtained by grinding and, if necessary, lapping to surface finishes in the range 0.10–0.15 μm CLA.

The two coatings most widely used are 60% Al_2O_3/40% TiO_2 and Cr_2O_3. Chromic oxide has the greater wear resistance and requires diamond wheels for grinding. However, because of its greater hardness it is possible to obtain smoother surfaces and therefore higher surface/yarn friction values. The softer alumina titania coating is more readily brush finished to a low friction value.

To protect against corrosion the coating is vacuum impregnated with a curable resin for low temperature applications. For applications such as yarn heater

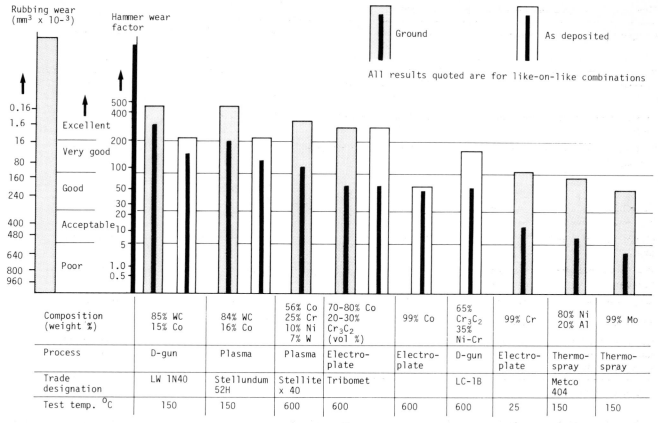

Composition (weight %)	85% WC 15% Co	84% WC 16% Co	56% Co 25% Cr 10% Ni 7% W	70-80% Co 20-30% Cr_3C_2 (vol %)	99% Co	65% Cr_3C_2 35% Ni-Cr	99% Cr	80% Ni 20% Al	99% Mo
Process	D-gun	Plasma	Plasma	Electro-plate	Electro-plate	D-gun	Electro-plate	Thermo-spray	Thermo-spray
Trade designation	LW 1N40	Stellundum 52H	Stellite x 40	Tribomet		LC-1B		Metco 404	
Test temp. °C	150	150	600	600	600	600	25	150	150

Fig. 3.10 Wear performance of electrodeposited and sprayed coatings.[21]

tracks, where the temperature is higher than that at which the resin decomposes, an undercoat of electro-less nickel can be used instead.

Fig. 3.11[23] shows sections through two types of brass heater tracks illustrating the configuration of the electroless nickel undercoat (~0.05 mm) and the chromium oxide (maximum thickness ~0.05 mm) sprayed coating.

Fig. 3.11 Section through yarn heater track.[23]

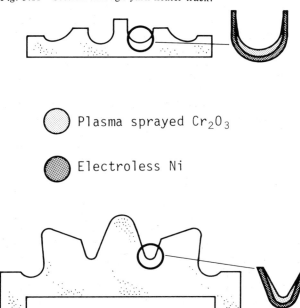

Plasma sprayed Cr_2O_3

Electroless Ni

7. Cemented carbide cutting tools

Metal cutting tools undergo wear by a number of mechanisms, eg diffusion, adhesion and abrasion. Diffusion wear, which is considered to be responsible for the cratering of the rake face of single point tools, is caused by the diffusion of the tool consituents into the chip material at the high temperatures generated at the cutting edge. Flank (or clearance face) wear is mainly the result of abrasion by the chip material. To improve tool performance, coating materials must therefore be harder than the tool material and less soluble in the chip material. In the case of cemented carbide tools potential coating materials are restricted to the harder carbides, nitrides and oxides. Methods available for depositing dense adherent coatings of these materials are chemical vapour deposition and physical vapour deposition processes, such as sputter-ing and ion plating. Chemical vapour deposition, having been developed considerably earlier than the physical vapour deposition processes, is now well established as the preferred cemented carbide coating process. It has the advantages over the PVD processes of superior throwing power and greater bond strengths because of coating-substrate interdiffusion at the processing temperature (~1000°C). The properties of cemented carbides are relatively unaffec-ted after exposure to this temperature. The coating of cemented carbide by PVD processes has neverthe-less been investigated but has not, so far, been adopted commercially.[24]

The materials most widely used for coating cemented carbide tools by CVD are TiC, TiN, TiCN (a mixed carbo-nitride) and Al_2O_3. The relative merits of these materials in resisting the various forms of wear are given in Table 3.10.[25] TiN and Al_2O_3, although less hard than TiC, resist cratering better. TiC, however, is more resistant to flank wear.

Table 3.10 Resistance of CVD coatings to various types of wear[25]

Mechanisms of Wear	TiC	TiCN	TiN	Al_2O_3
Diffusion	+ +	+ +	+ + +	+ + +
Adhesion	+ + +	+ +	+	+ +
Abrasion	+ +	+ +	+	+ + +
Oxidation	+	+ +	+ + +	+ + +

Key: + Medium   + + Good   + + + Excellent

A large amount of information has been published on the effects of various coatings on the performance of cemented carbide cutting tools. For example Fig. 3.12 shows tool life as a function of cutting speed in the turning of 1045 steel, using uncoated, TiC- and Al_2O_3-coated inserts.[26] Life decreases rapidly with increasing speed in all cases but for a life, say, of 10 min. a TiC coating allows a 50% increase in speed and an Al_2O_3 coating, a 90% increase. The curves cross over broadly because hardness is the controlling factor at low speeds and chemical stability at higher speeds. The low thermal conductivity of Al_2O_3 is also thought to be a factor at high cutting speeds as it acts as a thermal barrier, helping to maintain the tool at a lower temperature. Fig. 3.13 presents similar curves, determined under slightly different conditions, and including data for a TiN coating; this behaves like TiC, but does not increase tool life quite as much.[27] Curves obtained on cast iron, as shown in Fig. 3.14, display the same pattern, but the cross-over points are at lower speeds, indicating that Al_2O_3 is better that TiN or TiC at most practical cutting speeds on cast iron. The curves in Figs. 3.15 and 3.16, showing time to a given degree of crater and flank or clearance face wear, respectively, as a function of coating thickness, illustrate that alumina is the more resistant to crater wear and

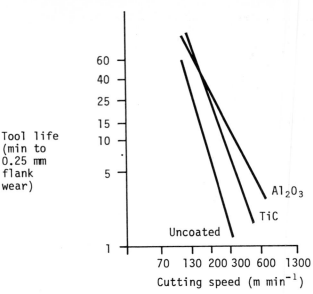

Fig. 3.12   Tool life as a function of cutting speed.[26] (AISI 1045 (190 HB) 0.25 mm/rev, depth of cut 2.5 mm)

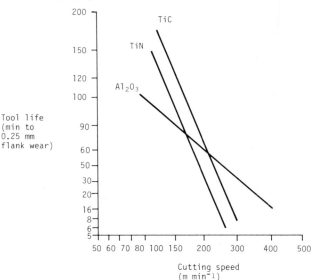

Fig. 3.13   Tool life as a function of cutting speed.[27] (AISI 1045 (190 HB) 0.4 mm/rev, depth of cut 2.5 mm).

Table 3.11   Machining results with different coatings[25]

Coating	Turning Steel	Milling Steel	Turning Cast Iron	Cast Iron
	Wear rate μm/min	Breaking feed mm/tooth	Wear rate μm/min.	Crater depth μm@16 min
5 μmAl$_2$O$_3$	12	0.3	5	3
6 μmTiC + 2 μmAl$_2$O$_3$	10	0.7	7	8
6 μmTiC	18	0.7		
0.5 μmAl$_2$O$_3$ + 2 μmTiN + 1 μmTiCN + 4 μmTiC	8	0.8		
0.35 μmAl$_2$O$_3$ + 2 μmTiN + 1 μmTiCN + 4 μm TiC + 2 μmAl$_2$O$_3$	6	0.9	4	2
6 TiC + + 2 μmAl$_2$O$_3$ + 1 μmTiC	10	0.7	5	3

Note: The substrate for all tests was ISO M15.
Turning steel: AISI 1045 (190HB), 240 m/min, 0.2 mm/rev, depth of cut 2.5 mm.
Milling steel: AISI 1060, 88 m/min, varying feed rate, depth of cut 3 mm.
Turning cast irons: Centrifugal cast iron, 130 m/min, 0.5 mm/rev, depth of cut 2.5 mm.

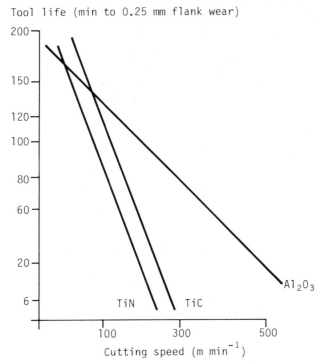

Fig. 3.14 Tool life as a function of cutting speed.[27]
(SAE G4000 cast iron, 210 HB, 0.25 mm/rev. depth of cut 2.5 mm)

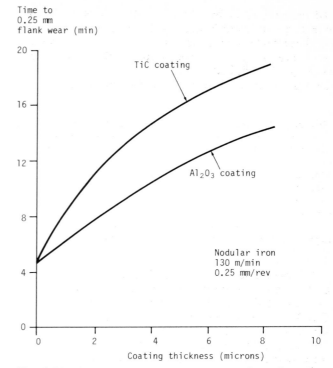

Fig. 3.16 Flank wear resistance as a function of coating thickness.[24]

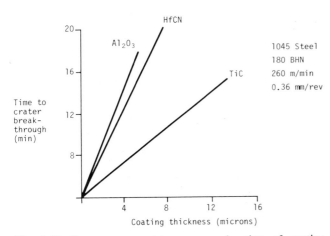

Fig. 3.15 Crater wear resistance as a function of coating thickness.[26]
(AISI 1045, 180 HB, 0.36 mm/rev, 260 m/min).

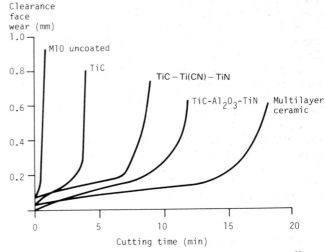

Fig. 3.17 Clearance face wear as a function of cutting time.[28]

titanium carbide to flank wear.[26]

Multilayer coatings can demonstrate considerably higher resistance to clearance face wear than a single layer of TiC, as illustrated in Fig. 3.17.[28] These tests were made on plain carbon steel, but similar benefits were observed on grey cast iron, the lifetime for the various multilayer coatings, under identical conditions, being: 8.5 min for TiC/Ti(C,N)/TiN, 12.5 min for TiC/Al₂O₃/TiN and 19 min for grade Sr 17 (a proprietary coating consisting of ten layers).[28] Some further results on multilayer coatings are repro-

duced in Table 3.11; the best performance being obtained with a multilayer coating that had Al₂O₃ at the top and bottom.[25] An initial layer of Al₂O₃ (especially with TiN above it) is believed to act as a barrier preventing decarburisation and consequent loss of edge strength. An outer layer of Al₂O₃, because of its low conductivity and chemical resistance, functions as an effective thermal and wear barrier.

CVD coatings have also been shown to reduce the forces required in cutting, the effect being greater at low cutting speeds.[28]

8. High speed steel metal cutting tools

High speed steels (eg M2 type) soften when heated above 700°C and therefore, when coated by chemical vapour deposition which requires higher temperatures, they must undergo a post coating heat treatment. Components also frequently distort. The CVD process is therefore less attractive for coating high speed steels than cemented carbides.

The advent of the PVD processes, in particular reactive ion plating in which the substrate temperature can be maintained below 500°C, has allowed the coating of high speed steel tools without softening or distortion. Some of the types of tools coated and the improvements in life gained are shown in Table 3.12.[29] The process is economically more attractive on the more complex tools, such as hobs and milling cutters where sharpening is relatively expensive, than on drills where sharpening is an inexpensive operation. Most of the information on tool performance is nevertheless on drills. Fig. 3.18 is a diagram of a drill tip showing the names given to the various surfaces. The performance of various coated drills are given in Table 3.13.[30]

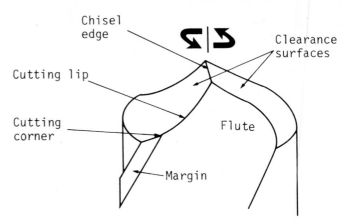

Fig. 3.18 Drill tip showing names given to various surfaces.[30]

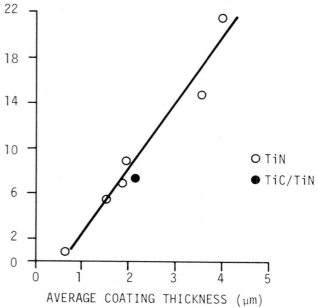

Fig. 3.19 Drill life improvement factor versus average coating thickness for PVD coated drills.[30]

Table 3.12 Improvements in life of PVD coated high speed steel cutting tools[29]

Tool	Coating	Improvement factor	Source
Tap	TiC	5–10	Japan
End-mill	TiN	3–8	Japan
Milling cutter	TiN	3	Japan
Hob	TiN	2–3	Japan
Gear cutter	TiN	2	Japan
Drill	TiN	4–8	Japan
Boring cutter	TiN	4–10	Japan
Drill	TiN, MoN or Mo$_2$N	2	USSR
Punches	TiN, MoN or Mo$_2$N	3	USSR
Broaches	TiN, MoN or Mo$_2$N	1.7	USSR

Table 3.13 Summary of drill test, coating thickness and substrate hardness[30]

Coating	No of Holes Drilled		Improvement Factor	Coating Thickness (µm)	Substrate Hardness
	Uncoated	Coated	(Coated/ Uncoated)	Clearance/Margin/ Flute	(HV–500 g)
UNCOATED					807 ± 27
PVD 1 (TiN)	93 ± 54(8)*	636 ± 140(8)	6.8	19/2.3/1.4	821 ± 50
PVD 2 (TiN)	84 ± 42(5)	749 ± 257(5)	8.9	4.0/1.1/0.7	735 ± 131
PVD 3 (TiC/TiN)	84 ± 42(5)	624 ± 82 (6)	7.4	2.2/2.3/2.0	737 ± 62
PVD 4 (TiN)	81 ± 13(3)	61 ± 8 (3)	0.8	0.7/0.7/0.6	799 ± 82
PVD 5 (TiN)	—	0 (3)	0	2.3/2.5/1.2	395 ± 33
PVD 6 (TiN)	84 ± 42(5)	464 ± 66 (2)	5.5	1.0/2.0/1.6	807 ± 34
PVD 7 (TiN)	55 ± 22(5)	793 ± 25 (2)	14.8	1.8/4.8/4.0	819 ± 56
PVD 8 (TiN)	32 ± 14(7)†	693 ± 157(2)†	21.7	2.1/5.4/4.4	800 ± 66
CVD 1 (TiC/TiN)	62 ± 10(5)	677 ± 135(5)	10.9	10.5/9.0/9.5	755 ± 50
CVD 2 (TiC/Ti (CN)/TiN)	54 ± 20(5)	620 ± 149(5)	11.5	5.1/5.0/5.2	860 ± 66
CVD 3 (TiC/Ti (CN))	57 ± 28(5)	47 ± 46 (5)	0.8	1.8/2.2/2.0	761 ± 75

*Number in parenthesis indicates the number of drills tested.
†Tested with harder steel plates, 36–39 HRC, (the rest with softer plates, 32–35 HRC)

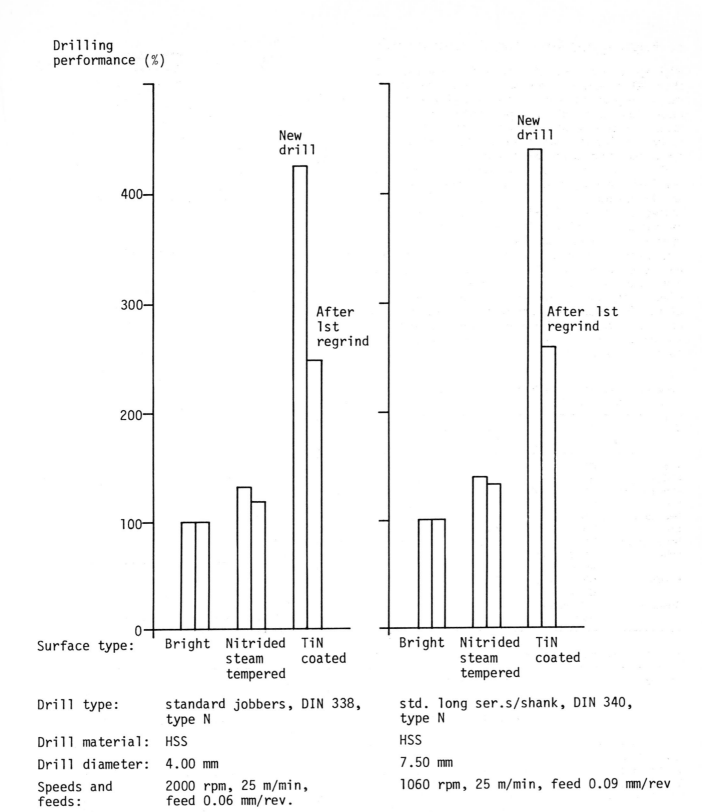

Drilling performance (%)

	New drill	
	After 1st regrind	

Surface type: Bright / Nitrided steam tempered / TiN coated

Drill type:	standard jobbers, DIN 338, type N	std. long ser.s/shank, DIN 340, type N
Drill material:	HSS	HSS
Drill diameter:	4.00 mm	7.50 mm
Speeds and feeds:	2000 rpm, 25 m/min, feed 0.06 mm/rev.	1060 rpm, 25 m/min, feed 0.09 mm/rev

Fig. 3.20 Effect of TiN coating or nitriding and steam tempering on the performance of two types of drill.[31]

Coatings of PVD 1–5 were produced by reactive ion plating and coatings 6–8 by biased reactive sputtering. The poor performance of PVD 5 is explained by the low substrate hardness and the relatively inferior performance of PVD 4 by the thinness of the coating.

The effect of PVD coating thickness on drill life is shown in Fig. 3.19.[30] This effect is not usually recognised by those reporting results on tool trials so that some of the conclusions drawn on the effect of other variables may be invalid. It would appear that for a given improvement in drill life a thicker CVD coating is required and this may explain the inferior performance of CVD 3.

In Fig. 3.20[31] the lives of treated and untreated drills when used to drill through holes in 15 mm alloy steel plate are compared. TiN coated drills perform much better than bright (untreated) or nitrided and steam tempered (a treatment which produces a black oxide layer which helps oil retention) drills, and retain a substantial proportion of the advantage after the first regrinding of the point.

Fig. 3.21 illustrates the variation of drill life with cutting speed when drilling 13 mm through holes in AISI 4340 steel. A TiN coating ensures a longer life at all speeds or alternatively can sustain a higher speed for a given lifetime. The benefit of a TiN coating on an end mill (M7 steel) is shown in Fig. 3.22. In this work it is claimed that for a given length of cut wear decreases with increasing cutting speed for coated tools.

The nature of the workpiece material has an important effect on the degree of improvement conferred by a TiN coating (Table 3.14). For any one material there is a wide spread in the improvement obtained

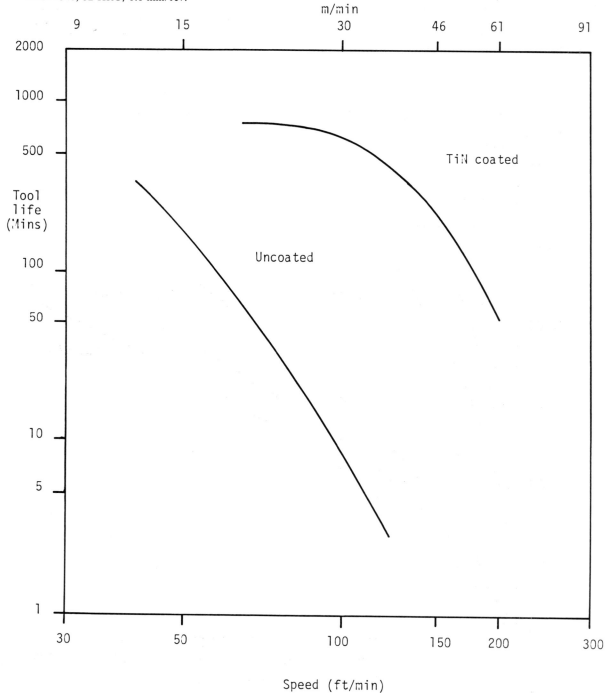

Fig. 3.21 Drill life as a function of cutting speed.[32]
AISI 4340, 32 HRC, 0.1 mm/rev.

Speed (ft/min)

Table 3.14 Percentage increase in number of holes produced in various materials using TiN coated drills[33]

Work material	Increase %		
	Min	Avg	Max
Case-hardening steel, cementation steel (5115, ASTM A369)	110	180	200
Low-carbon steel, heat-treatable to 115,000 psi (1020, 1043)	268	327	400
Med-carbon steel, heat-treatable to 145,000 psi (1045, 1060)	0	183	430
Alloy steel, heat-treatable to 190,000 psi (6150, O2, D7)	0	142	500
Structural steel (60, 75)	85	269	500
Grey iron	15	269	800
Free-machining steel (1112)	100	290	420
Tool steels (D6, 6F2)	150	380	1200
Aluminium-silicon alloy	100	424	1330
Spheroidal-graphite iron	590	710	1330
Cold-extrusion steel (1030)	420	912	1800
Stainless, heat-resistant steels (304, 316, 440C)	0	923	11,166
Bronze alloy	650	775	900

Note: Alloy designations in parentheses are approximate US equivalents

Fig. 3.22 Wear of end mill at different cutting speeds.[3] (Cutting speeds in surface feet per minute, sfm).

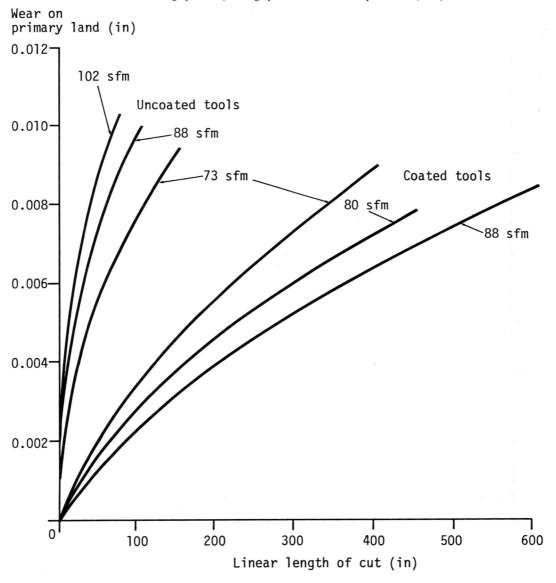

but nevertheless the overall performance is impressive. It is concluded that coatings are especially advantageous for the more abrasive materials and those that produce short chips.

9. Metal forming tools

Metal forming tools are subjected to severe abrasive wear by sliding against the metal being formed. They are also subjected to impact forces which can give failure by fatigue. In the case of hot forging tools, thermal fatigue can also cause cracking. A high surface hardness is desirable to resist abrasive wear and any surface treatment must give a layer integral with the substrate or sufficiently well bonded to resist the imposed impact forces and thermal cycles. Nitriding of high speed (eg M2) and cold work (eg D2) steels for cold working tools such as dies and punches is used extensively. Because the process is carried out at a relatively low temperature there is little distortion of the tool and high surface hardnesses are obtained because of the high concentration of nitride forming elements (Cr, Mo, V) in these steels.[34] An example is the cold forming dies used in the manufacture of car drive axle couplings.[35] To improve the performance of hardened and tempered dies, which suffer pick up and chipping of the working faces during operation, a number of surface treatments were investigated (Table 3.15). A standard nitrocarburising treatment

Table 3.15 Performance of surface treated cold forming dies[35]

Material	Treatment	Number of components produced (Average)
BM2 or ASP23	Hardened and tempered.	2,000
	Hardened and tempered, nitrocarburised	4,000
	Hardened and tempered, plasma nitrided	20,000
	Hardened and tempered, CVD TiN/ TiC coated; re-hardened and tempered.	50,000

(Tufftriding) doubled the number of components produced by each tool and further improvements were obtained using plasma nitriding when a two hour cycle at 510°C in cracked ammonia increased the surface hardness to 1500 HV. More recently CVD has been used to produce TiC/TiN coatings. After this high temperature treatment the die is rehardened and tempered in a high pressure gas quench vacuum furnace. A coating ~9 μm in thickness, consisting of an inner layer of TiC (~3000 HV) and an outer layer of TiN (~2500 HV), has produced a substantial improvement in tool life. The cost savings in enhanced tooling life and reduction in tooling change time justify the increase in treatment costs from 20 p/kg for hardening and tempering to £2.20 p/kg for CVD coating.

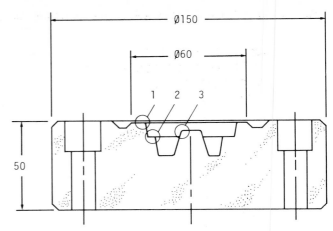

Fig. 3.23 Hot forging die in 0.3%C, 3%Cr, 3%MoV steel.[26]

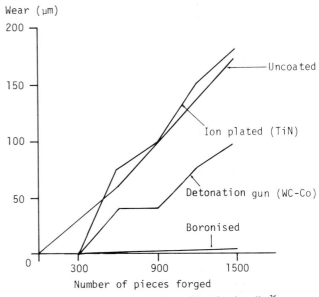

Fig. 3.24 Wear on boss nose radius of hot forging die.[36]

High surface hardnesses are also obtained by boriding (~2000 HV) and by metalliding (~3500 HV) and both these processes are finding application in metal forming tools.[34] Boriding has been used[35] on the hot forging dies shown in Fig. 3.23. Blanks in 0.3% C, 1.0% Cr steel were heated to 1100°C and forged in one blow, the pressure required being 600 N/mm². The wear on radius 3 of the diameter was measured after every 300 forgings. The results shown in Fig. 3.24 are compared with other surface treatments, detonation gun coating of WC-Co and ion nitrided TiCN. This latter coating presumably failed in the early part of the test, as the wear rate reverted to that obtained on the uncoated tool. Considerable improvements have been made in the quality of ion plated coatings since this work was reported (1977) so that better results would now be expected.[37] Fig. 3.24 shows the almost zero wear rate of the borided dies.

The treatment of metal forming dies by the Toyota Diffusion (TD) process is expanding rapidly in Japan (Fig. 3.25). The increases in life of various types of

Fig. 3.25 Growth in use of TD process on tools and dies in Japan.[38]

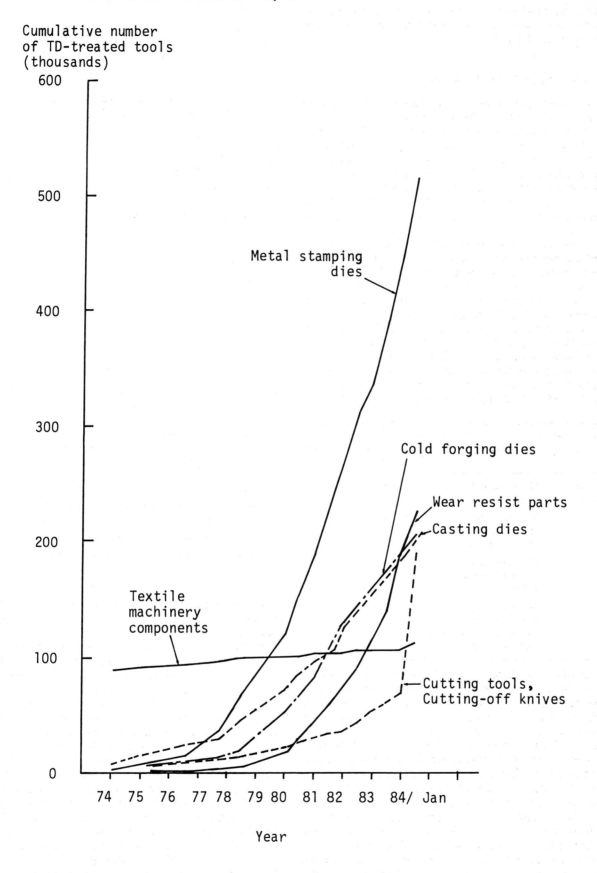

Cumulative number
of TD-treated tools
(thousands)

600 —

500 —

400 —

300 —

200 —

100 —

0 —

Metal stamping
dies

Cold forging dies

Wear resist parts

Casting dies

Textile
machinery
components

Cutting tools,
Cutting-off knives

74 75 76 77 78 79 80 81 82 83 84/ Jan

Year

Table 3.16 Typical life improvements achieved by TD processing of tools and dies[38]

Type of tooling	Life increase factor	Life increase Cost increase
Piercing punch (cold work steel)	5–10	2–8
Cold forging punch (HSS)	5–10	2.6–5.2
Carbide cold forging punch replaced by HSS with VC coating	1	5.6
Core pin for aluminium gravity diecasting using H series die	3–5	2.3–3.8
Hot-forging dies	2–5	–
Extrusion dies	2–15	–
Steel powder compaction	10–30	–
Plastic moulding	5–10	–
Aluminium diecasting	2–10	–

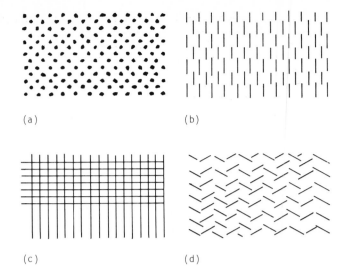

Fig. 3.26 Spreader techniques for weld deposits.

tools by using the treatment are shown in Table 3.16.[38]

The main limitation of CVD, boriding and the TD process is the distortion that occurs during processing at around 1000°C and subsequent heat treatment. Grinding and lapping of such hard surfaces is a slow and expensive operation so that only parts with dimensional tolerances of greater than 20 μm can be treated. This usually restricts the application of these processes to relatively small parts.

10. Rock, ore and earth engaging equipment

This type of equipment is subjected to low stress abrasion and grinding abrasion by relatively fine particles of soil, sand and ore and gouging abrasion by rocks and larger lumps of ore. The required precision in the dimensions of components is usually low and a substantial amount of wear (several centimetres) can usually be tolerated. In addition to being hard, materials used must be sufficiently tough to resist the impact forces associated with gouging wear and any surface treatment must be capable of providing thick layers well bonded to the substrate. Components are therefore usually made from tough low carbon martensitic steels which may in addition be surface treated by induction hardening, carburising or hard facing. Thus bulldozer scraper blades, made in a weldable boron steel (eg AISI 86B30H) are induction hardened. Rotating tooth cones of rolling cutter rock drilling bits, made in nickel-molybdenum alloy steel (eg. AISI 4620) are carburised on the inner bearing surfaces and hardfaced on the outer surfaces with tungsten carbide in an iron base.

Wear rates on rock and earth engaging tools and equipment are nevertheless high and necessitate the observance of strict maintenance schedules. The high cost of carrying large numbers of spares means that mine or quarry operators tend, as far as possible, to reclaim worn parts, usually by welding. This may be carried out by re-building the worn surfaces, either with weld metal or by welding on suitable wear plates or bars, following which the new surfaces may be finished by hardfacing.

The range of parts reclaimed by hardfacing is extensive and includes: tractor rollers and tracks; dragline and excavator buckets and teeth; crusher hammers and rolls; mixer blades; dredger blades and ore chutes. The hardfacing materials employed range from low alloy steels for build-up operation, austenitic manganese steels for parts subject to heavy impact from rock or ore, martensitic and high speed steels for non-lubricated metal-to-metal wear, and high chromium irons or tungsten carbide composites for very abrasive conditions. Typical hardfacing schedules for earth moving equipment are given in Table 3.17[39] and illustrate how different hardfacing materials are used to resist varying types of abrasive wear.[40] In crushing equipment, where gouging and impact are dominant, austenitic manganese steel is widely used both in original equipment and as hardfacing for repairs.

Considerable economies in hardfacing alloys, together with improved component performance, are secured by depositing the weld metal in patterns, as illustrated in Fig. 3.26. These patterns can often be made without necessarily breaking the arc during the deposition process.

The pattern most appropriate to a given wear situation is dependent upon the type of abrasive material handled and on its flow path.[19] Thus in the case of a digger tooth handling rock it may be better to apply the hardfacing in stringer beads running parallel to the direction of flow, as shown in Fig. 3.27 (a); this helps to prevent undue wear on the beads themselves, since these tend to channel the flow of rock between them. On the other hand, for teeth handling sand it may be better to apply the beads at right angles to the direction of travel, again leaving spaces between the beads, as shown in Fig. 3.27 (b). Sand will then become trapped between the beads and this will to some extent protect the underlying metal. Compromise patterns can be employed where mixed rock and sand are being handled (Fig. 3.27 (c)).

A special case is the hardfacing of cutting, drilling or digging tools where a sharp edge must be maintained. In such cases it is usual practice to face one side of the tool only, so that as the other side wears away in service, a self sharpening edge is produced (Fig. 3.28). With thin cutting edges care must be

Table 3.17 Typical hardfacing schedule for earth moving equipment[39]

Component	Base Metal	Hardfacing
Track components (rollers, idlers, top carrier rolls drive sprocket, tractor rail)	Carbon steel	3 layers of low carbon martensitic steel giving a deposit hardness of 360–470 HV.
Ripper teeth	Carbon steel or austenitic manganese steel	2 layers of austenitic high chromium iron giving a deposit hardness of 653–746 HV.
Dozer end bit	Carbon steel	Continuous layers of tungsen carbide particles in iron matrix (particle size of WC related to application) on edge, on remainder use latticed stringers of martensitic steel giving a deposit hardness of 577–653 VPN.
Shovel teeth	Austenitic manganese steel or carbon steel	Hardfacing pattern, as well as hardfacing material, important in controlling wear.
Shovel bucket	Carbon steel	Materials as for shovel teeth. Continuous coating on lips and inside base—lattice on remainder.

For the Shovel teeth row, additional detail:

Sand or soil (abrasive)	*Rock or stone* (heavy impact)
High chromium irons	Martensitic steel with or without
Tungsten carbide in iron base.	tungsten or titanium. Cover whole face or use stringer beads parallel to flow.

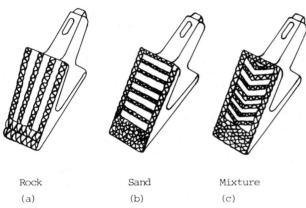

Rock   Sand   Mixture
(a)   (b)   (c)

Fig. 3.27 Controlling flow of material.

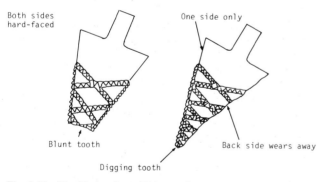

Both sides hard-faced   One side only

Blunt tooth   Back side wears away

Digging tooth

Fig. 3.28 Hardfacing for self sharpening.

exercised to avoid overheating and distortion, and for this reason hardfacing of these parts is generally carried out using the oxy-acetylene process.

Determination of the optimum deposition of hardfacing alloy is often made easier if the wear patterns arising in service can be observed, so that the hardfacing may be applied where it is most needed. This matching of hardfacing to a wear pattern is illustrated in Fig. 3.29.

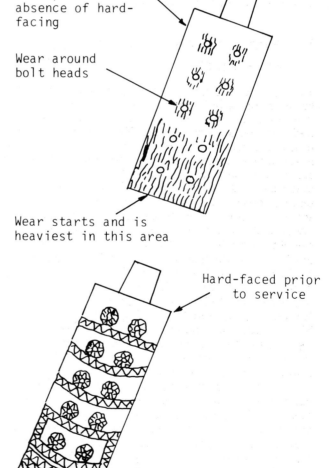

Wear pattern in absence of hard-facing

Wear around bolt heads

Wear starts and is heaviest in this area

Hard-faced prior to service

Fig. 3.29 Applying hardfacing to match wear pattern.

43

Most reclamation work on site is done using manual metal arc welding, although MIG and open arc welding using flux-cored wires may be employed where the volume of work warrants this. Fully automatic methods are rarely used, except by firms specialising in reclamation work, who may for instance use submerged arc welding for resurfacing worn track rollers. A number of quarries in the UK do however employ fully automatic equipment for in-situ resurfacing of rotors in rock crushing hammers, a procedure that avoids exposure of the welder to the effects of heat and fumes generated in the enclosed crushing chambers; it also increases productivity by at least 150%.

It has been stated that hardfacing costs represent between 25% and 40% of mining companies' total repair and maintenance budgets which, in open cast mining, may amount to 40–50% of direct costs.

An alternative to hardfacing on-site is the use of wear plates, strips or buttons which are made of steel already clad with high chromium austenitic iron and are readily fixed in place by welding the steel backing directly to the worn component.

Soil engaging tools used in farm tillage, such as ploughshares, ploughsweeps and ploughdiscs, have to resist the abrasive wear caused by silica particles in the soil. Traditionally these tools are made in 0.4%C carbon-manganese, 0.7%C spring steel or 0.8%C carbon-silicon-manganese spring steel. When heat-treated to a hardness of about 500 HV these materials give the best compromise between wear, impact and fatigue resistance. Tests have shown that the best hardfacings are austenitic and martensitic irons and that correct orientation of the weld deposit with respect to the soil flow improves the wear rate.[41, 42] In practice the variable nature of soils in different parts of the country, in particular the size of stones, makes it difficult to generalise on the economic benefits of hardfacing of tillage tools.

In soil of low stone content it is possible to use solid ceramic (high alumina) facings on sub-soilers, cultivator blades and mole expanders.

11. Chute liners for coke and blast furnace sinter

In these applications the main wear mechanism is low stress abrasion and the primary material requirement is that it is harder than the abrasive. However, where the abrasive falls onto the chute there are also impact loads which are larger with higher density materials such as iron ore sinter. A high degree of wear is tolerable and solid plates of steel (eg 13% Mn) or cast iron or tiles of ceramics, such as alumina or cast basalt, are frequently a more economic solution than hardfacing.

The behaviour of abrasion resistant material as chute liners subjected to:

(a) sliding wear by coke,
(b) sliding wear by blast furnace sinter, or
(c) impact wear by blast furnace sinter (free fall $\simeq$ 1.25 m)

was investigated by Hocke;[43, 44] the results are reported in Tables 3.18–3.20 as a wear index, ie the wear rate of the test material relative to that of a 0.40% C/0.5% Mn steel (En 8) under similar conditions. Ceramic materials performed better when subjected to sliding wear by coke than by sinter and this was thought to be due to the greater impact loads with the higher density sinter (about four times that of coke).

Table 3.18 Wear indices of wear resistant materials subjected to sliding wear by coke[43]

Material	Wear index*
Sintered tungsten carbide	0.028
Fusion-cast alumina	0.053–0.115
High-chrome hardfacing	0.207
High-chrome martensitic white cast iron	0.231–0.328
Ni-Cr martensitic white cast iron	0.240
Cr-Mn-Mo hardfacing	0.243
13% Mn skin-hardened steel plate	0.307
Slagceram	0.326
$3\frac{1}{4}$% Cr-Mo cast steel	0.370
Cast basalt	0.378
Acid-resisting ceramic tile	0.413
Iron-based chrome alloy, cast	0.427
Low-alloy steel plate, quenched and tempered	0.460–0.665
Nodular graphite-based cast iron	0.470
13% Mn austenitic cast steel	0.476
Low-alloy cast iron	0.535
High-phoshorus pig iron	0.700
Concrete	0.910–1.900
$1\frac{1}{2}$% Cr-Mo cast steel	0.920
Plate glass	1.760
Solid rubber	1.850–4.800
Quarry floor tiles	4.780–7.350
Polyurethane	4.930–11.630
Resin-based calcined bauxite	5.000
Profiled rubber	6.880
High-density polyethylene	14.000
Polytetrafluorethylene	17.800

*Volume lost relative to that of 0.4% carbon steel (En 8).

Table 3.19 Wear indices of wear resistant materials subjected to sliding wear by blast furnace sinter[43]

Material	Wear index*
Ni-Cr martensitic white cast iron	0.065
Cr-Mo-W-Nb hardfacing	0.079
High-chrome martensitic white cast iron	0.115
Nodular graphite-based cast iron	0.137
Fusion-cast alumina special	0.172
Sintered alumina	0.173
Tungsten carbide hardfacing	0.192
Fusion-cast alumina	0.221
Low-alloy steel plate, quenched and tempered	0.382–1.336
Slagceram	0.518
Low-alloy cast iron	0.074
Fusion-cast basalt	0.846
High-phosphorus pig iron	0.953
Acid-resisting ceramic tile	2.030
Silicon carbide ceramic	2.265–8.470
Polyurethane	2.490
Concrete	6.350–18.00

*Volume lost relative to that of 0.4% carbon steel (En 8).

Table 3.20 Wear indices of wear resistant materials subjected to medium impact abrasion by blast furnace sinter[43]

Material	Wear index*
Cr-Ni-Mo martensitic white cast iron	0.144
Ni-Cr martensitic white cast iron	0.157
High-chrome martensitic white cast iron	0.158
Nodular graphite-based cast iron	0.274
High-chrome hardfacing	0.295
Fusion-cast alumina	0.323
Extruded alumina	0.350
13% Mn cast steel	0.536
Low-alloy cast iron	0.595
Low-alloy steel plate, quenched and tempered	0.750

*Volume lost relative to that of 0.4% carbon steel (En 8)

The data is thought to be equally valid for feeders, hoppers, screen mats and blast furnace ancilliaries such as skips, charging hoppers and bells.

12. Pulverised fuel handling equipment

Considerable abrasive wear occurs in pulverising coal for power station boilers, particularly when the silica content of the coal is high. Exhauster fan blades in suction grinding mills are subjected to severe erosive wear because of the high fan speed. Various surface coatings have been tried[45] but only sintered carbide plates have given acceptable lives (Table 3.21). The plates, which are 3 mm thick and hexagonally shaped (2.5 cm side), are brazed onto the mild steel fan blades (Fig. 3.30), the bond strength being critical

Table 3.21 Life of mild steel exhauster fan blades in coal pulverising mill handling high silica content coal[45]

Coating	Thickness, mm	Life, h	Remarks
Nil	—	300/400	
Plasma sprayed alumina on surface, Stellite on leading edge	0.75	539	
Sprayed molybdenum	0.25	476	
	0.75	628	
Chromium boride paste	0.80	>572	Test stopped prematurely
70% tungsten carbide flame sprayed: powder process	0.75	1100	Blade carriers collapsed
Tungsten carbide platelets (on part of blade only)	—	661	Platelets satisfactory, unprotected areas worn away
Tungsten carbide platelets	—	1800	Wear between platelets
	—	2800	Some platelets detached
Tungsten carbide angle tiles/platelets	—	4800/5000	Recommended treatment
Alumina tiles	12.5	624	Heavy erosion on leading edge

Table 3.22 Examples of improvements in life obtained by surfacing components in coal pulverising mills[45]

Designation	Original part		Surfaced part		Life improvement factor
	Material	Life, h	Material	Life, h	
Mill body (joint area)	Cast iron	4000	Tungsten carbide platelets on mild steel	>23,000	>5.75
Wear plate	25 mm thick mild steel	3000	Tungsten carbide platelets on mild steel	>23,000	>7.6
Liner plate	Mild steel	3000	Molybdenum sprayed	6000	2
Dam ring	10 mm thick mild steel	1500	Chromium boride on 22 mm thick mild steel	6000	4
Wear plate and baffle	Mild steel	1400/1600	Chromium boride coated	6000	4
Separator blades	Mild steel	3000	Chromium boride coated	6000	2
Separator liner plates	Not specified	15,000	Molybdenum sprayed	>18,000	>1.2
				30,000*	2*
Shaft journal (mating with seal)	Not specified	20,000	Tungsten carbide hard faced	50,000*	2.5*

*Estimated lives

since the joint has to sustain forces in excess of 14 G on start-up. The cost of using tungsten carbide tiles is high, about 3.5 times that of other forms of hardfacing.

Improvements in component life by surface coating other parts of a grinding mill are shown in Table 3.22. The use of these techniques allows the lives of components to be extended to 6000 h or more, after which time the Ni-hard grinding rollers require refurbishing.

Fused basalt linings are widely used in mild steel pipelines carrying pulverised fuel suspended in air. Erosion tends however to occur at bends, the softer glassy phase wearing away from around the harder particles in the basalt. Alumina tiles or chromium boride coatings are therefore used in pipe bends. Fused basalt is also used for lining sluiceways carrying wet pulverised fuel ash, the wear process being low stress abrasion rather than particle impact erosion.

Fig. 3.30 Tungsten carbide tiles applied to exhauster fan blades.

References

[1] *Machine cut gears, A. Helical and straight spur*. British Standard 436.

[2] NEALE, M J *et al*. Gears, Section 1c–2b3 of *Fulmer Materials Optimiser*, Fulmer Research Institute Ltd, Stoke Poges, 1974.

[3] TUPLIN, W A. Gears, selection of type and materials. Section A24 of *Tribology Handbook*, Butterworths, London, 1973.

[4] DUDLEY, D W. Gear wear. *Wear Control Handbook*, ed Peterson, M B and WINER, W O. ASME, NY p 755, 1980.

[5] FOSBERRY, R A C and MANSION, H D. Durability of gears: the effect of materials and heat treatment on gear scuffing. *Motor Industry Research Association Report*, Nuneaton, No 1949/8.

[6] KOROTCHENKO, V and BELL, T. Applications of plasma nitriding in UK manufacturing industries. *Heat Treatment of Metals*, Vol 5 (4), p 88, 1978.

[7] WYDLER, R. Gear scuffing: state of calculation methods. *Forschungsstelle für Zahnräder und Getriebebau Colloquium*, Munich, p 147, 1973.

[8] MILLER, J E and WINEMAN, J A. Laser hardening at Saginaw Steering Gear. *Metal Progress*, 111, p 38, 1977.

[9] DAY, R A. Piston ring and liner wear. *Industrial Lubrication and Tribology*, p 44, March/April 1982.

[10] MULDER, E H. Chromium plating of medium speed diesel engine cylinder liners. *Diesel and Gas Turbine Worldwide*, p 12, 1980.

[11] MURRAY, E J. Piston rings for passenger car turbines. *SAE Paper* 790698, Detroit, 1979.

[12] TAYLOR, B J and EYRE, T S. A review of piston ring and cylinder liner materials. *Tribology International* Vol 12, 2, p 79, 1979.

[13] HYDE, G F, CROMWELL, J E and BARNES, J A. Piston ring coatings for internal combustion engines. *SAE Paper* 790865, Detroit, 1979.

[14] WELLWORTHY LTD. British Patent 1,320,902. 1969.

[15] DUECK, G E and NEWMAN, B A. Piston ring development—trends in Europe for off-highway applications. *SAE Paper* 810934, Detroit, 1981.

[16] COWLEY, W E, *et al*. Internal combustion engine poppet valves: a study of mechanical and metallurgical requirements *Proc. Inst. Mech. Engrs*, Vol 179, Pt 24, No 5, p 145, 1964–65.

[17] WEDGE, R H and EAVES, A V. Coatings in the aero gas turbine, *9th International Thermal Spraying Conf*. The Hague, p 73, 1980.

[18] THIEMANN, K and MALIK, M. Thermal spraying in transport aircraft. *9th International Thermal Spraying Conf*. The Hague, p 62, 1980.

[19] MEETHAM, G W. *The Development of Gas Turbine Materials*. Applied Science Publishers, London, 1981.

[20] TUCKER, R C. Plasma and detonation gun deposition techniques and coating properties. *Deposition Technologies for Films and Coatings*, ed Bunshah, R F. Noyes Publications NJ, p 454, 1982.

[21] KEDWARD, E C, WRIGHT, K W and TENNENT, A B. The development of electrodeposited composites for use as wear control coatings on aero engines. *Tribology International*. Vol 7, (5), p 221, 1974.

[22] GALLANT, P E, Ceramic coatings for the synthetic fibre industry, *Surfacing Journal*, Vol 12, (1), 1981.

[23] LAVIN, P A. British Patent 1,597,645. 1978.

[24] DODOMA, M, SHABAIK, A H and BUNSHAH, R F. Machining evaluation of cemented carbide tools coated with HfN and TiC by the activated reactive evaporation process. *Thin Solid Films*, Vol 54, 353, 1978.

25 PORAT, R. Thermal properties of coating materials and their effect on the efficiency of coated cutting tools. *Proc. 8th Int. CVD Conference Electrochemical Soc.*, 452, 1981.

26 GRAHAM, D E and HALL, T E. Coated cutting tools. *The Carbide and Tool Jnl*, 34, May–June 1982.

27 KALISH, H S. Status report: cutting tool materials. *Metal Progress*, Vol 124 (6), p 21, November 1983.

28 SCHINTLMEISTER, W, WALLGRAM, W and KANZ, J. Properties, applications and manufacture of wear resistant hard material coatings for tools. *Thin Solid Films*, Vol 107, p 117, 1983.

29 BUNSHAH, R F. Hard coatings for wear resistance by physical vapour deposition processes. *SAMPE Quarterly*, Vol 12, (1), 1980.

30 YOUNG, C T, BECKER, P C and RHEE, S K. Performance evaluation of TiN and TiC/TiN coated drills. *Conf. on Wear Materials*. Reston Va, 1982. p 235, 1982.

31 *Modern Twist Drill Technology*. Guhring Vertriebsgesellschaft, Albstadt, Vol 1, (1), 9, 1981.

32 HENDERER, W E. Performance of titanium nitride coated high speed drills. *N. Amer. Metalworking Res. Conf.* University of Wisconsin—Madison, 24–26 May, 1983.

33 HATSCHEK, R L, Coatings: revolution in HSS tools. *American Machinist*, p 129, March 1983.

34 CHILD, H C, PLUMB, S A and REEVES G. Influence of surface thermochemical treatments on properties of tool steels. in *Towards Improved Performance of Tool Materials, Book No 278*, p 135. The Metals Society, London, 1982.

35 STAINES, A M and BELL, T. Technological importance of plasma induced nitrided and carburised layers on steel. *Thin Solid Films* Vol 86, p 201, 1981.

36 FICHTL, W. Boronising and its practical applications. *Materials in Engineering*, Vol 2, p 276, December 1981.

37 Surface treatment of tools and dies—the options. Report in *Heat Treatment of Metals*, Vol 10, (3), p 77, 1983.

38 *TD Applications Literature*. Mitsui and Co Ltd, Temple Court, 11 Queen Victoria Street, London EC4N 4SB.

39 *Welding consumables for hardfacing*. Literature form Bohler Schweisstechnik.

40 FARMER, H B. Factors influencing selection and performance of hardfacing alloys. *Symposium on Materials for the Mining Industry*, Climax Molybdenum Co, Vail, Colorado, 1974.

41 FOLEY, A G. Reducing abrasive wear of soil engaging components in agriculture. *Welding Institute Seminar*, Coventry 1983.

42 MOORE, M A. Hardfacing soil engaging components. *The Agricultural Engineer* Vol 34, (1), 1979.

43 HOCKE, H. Wear resistant materials for plant handling coke and sinter *BSC Report* PE/B/5/72.

44 HOCKE, H. Wear resistant lining for steelworks chutes. *Tribology in Iron and Steelworks*. ISI London, Publ 125, 1970.

45 LEIVERS, D, NEI—International Combustion Limited. Derby. *Private Communication*.

CHAPTER 4
Thermal hardening

1. Introduction

In thermal hardening the surface of suitable materials, usually plain carbon or low alloy steels or cast irons, are austenitised and then quenched to produce a hard martensitic case which is usually tempered in a subsequent operation. Case depths are normally in the range 0.5–5 mm. Case hardnesses are typically 700 HV on hardening and 600 HV after tempering at 200°C.

Heating processes used include electrical induction and resistance, and direct impingement methods using flames, lasers and electron beams. Of these, induction heating is the most widely used. Laser and electron beam heating have recently become established in a number of applications mainly where distortion was a problem with the induction method.

Thermal treatments are mainly employed when only localised regions of engineering components require hardening. Consequently, such methods are energy efficient relative to thermochemical methods where bulk heating of batches of components is undertaken.

Induction surface hardening is applicable to axisymmetric or near axisymmetric components in steel or cast iron which are being produced in substantial volumes. Normally, the hardening process also introduces compressive stresses into the surface layers, leading to an improvement in fatigue properties. For example, the drive shafts of heavy lorries and buses are induction surface hardened to improve their fatigue properties.

The skills and experience required for manual flame hardening have been largely superseded by automated flame techniques or by the induction method. However, for one-off components the simplicity of manual flame hardening and the extremely low capital investment ensure that the method is still used on a regular basis.

From the heat treatment point of view the laser can be considered as a versatile and flexible high intensity heat source that can operate in air. It is capable of undertaking a range of processes, essentially simultaneously, since the laser beam can be directed through air by metal mirrors and switched and shared between a number of work stations. Manipulative techniques using mirrors allow the beam to be directed to areas not accessible by other techniques, eg the bores of tubes. Set against these considerations are the high capital cost and low energy efficiency of the technique.

2. Induction surface hardening

For induction hardening and most other thermal hardening processes suitable steels are plain carbon or low alloy steels with between 0.35 and 0.6%C. Below 0.35%C the treated surface is insufficiently hard and above 0.6%C there is a danger of surface cracking on quenching. To obtain a satisfactory hardening response on induction hardening it is necessary that the carbides dissolve when the steel is heated. Thus, steels that have been quenched and tempered and in which the cabide particles are small respond better than steels with large spherodised carbides. Similarly the rate of solution of carbides in

Table 4.1 Some steels which have been induction hardened[1]

En No.	C%	Mn%	Cr%	Ni%	Mo%	Nearest equivalent in revised BS 970
8D	0.40/0.45	0.70/0.90	—	—	—	080A42
9	0.50/0.60	0.50/0.80	—	—	—	—
10	0.50/0.60	0.50/0.80	—	0.50/0.80	—	—
11	0.50/0.70	0.50/0.80	0.50/0.80	—	—	526M60
12	0.30/0.45	1.50 max	—	0.60/1.00	—	503M40
15B	0.35/0.40	1.10/1.30	—	—	—	120M36
16C	0.35/0.40	1.30/1.80	—	—	0.20/0.35	605A37
18	0.35/0.45	0.60/0.95	0.85/1.15	—	—	530M40
19	0.35/0.45	0.50/0.80	0.90/1.50	—	0.20/0.40	709M40
24	0.35/0.45	0.45/0.70	0.90/1.40	1.30/1.80	0.20/0.35	817M40
43B	0.45/0.50	0.70/1.00	—	—	—	080A47
100	0.35/0.45	1.20/1.50	0.30/0.60	0.50/1.00	0.15/0.25	945M38

steels other than plain carbon and low alloy steels is too slow for them to be satisfactorily induction hardened. Steels must also have sufficient hardenability to achieve the required hardness at the specified case depth. Typical steels suitable for induction hardening are listed in Table 4.1.

In the induction hardening process an ac current flowing through an inductor (the work coil) causes eddy currents to be induced in the workpiece which result in heating. High frequencies, 500 kHz, are necessary for shallow cases about 0.5 mm deep, and low frequencies, 1 kHz, for depths of about 5 mm (Table 4.2).[1]

Table 4.2 Conditions for induction hardening[1]

Depth of hardening		Frequency kHz	Power input	
mm	(ins)		W/mm²	(kW/in²)
0.5/1.1	(0.020/0.045)	450	15–19	(10–12)
1.1/2.3	(0.045/0.090)	450	8–12	(5–8)
1.5/2.3	(0.06/0.09)	10	15–25	(10–16)
2.3/3.0	(0.09/0.12)	10	15–23	(10–16)
3.0/4.0	(0.12/0.16)	10	15–22	(10–14)
2.3/3.0	(0.09/0.12)	3	23–26	(15–17)
3.0/4.0	(0.12/0.16)	3	22–25	(14–16)
4.0/5.0	(0.16/0.20)	3	15–22	(10–14)

There are two basic techniques for induction hardening components, 'single-shot' and 'scanning'.[2] The former employs selective heating and quenching to harden a specific area or areas of the component in one operation. The latter is usually applied to harden progressively long, continuous sections, such as shafts and spindles. In this instance the scanning inductor traverses the length of the section, heating only a relatively small area at any given time, and is followed closely by the quench arrangement which is often an integral part of the inductor.

The advantages of electrical heating, in particular the induction method, are rapidity, repeatability and cleanliness. In addition, because induction heating is rapid and the hardening process times short, a batch process can often be replaced by a continuous one. Also, since the heat source is easily controlled, the hardening plant can be readily integrated into an automatic manufacturing production line.

3. High-frequency resistance hardening

The basic principle of this approach to selective surface hardening is illustrated in Fig. 4.1.[3] A water cooled 'proximity conductor' is placed close to the surface to be heated and connected to the workpiece, through a pair of contacts at the outer edges, and to a power source of typically 400 kHz. When the high-frequency current is applied, heating occurs across the work surface immediately beneath the proximity conductor. Once the heated strip reaches the hardening temperature, the power is switched off and self-

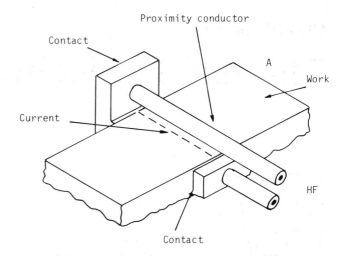

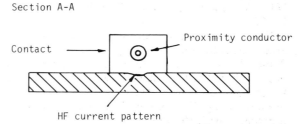

Fig. 4.1 Basic elements of selective surface hardening by high-frequency resistance heating.[3]

quenching by heat dissipation to the surrounding bulk occurs. The depth of hardening depends upon the frequency used, the time of heating and the power level, but is normally in the range 0.37–0.75 mm. A significant advantage of high-frequency resistance heating is that it does not require a closed loop of current but can heat a path between two points in the form of a terminated line, the pattern between two points being dictated by the shape of the conductor. With this rapid heating process, a very high power density can be achieved, with the result that a typical hardening cycle is usually less than 0.5 seconds. For example, on AISI 1095 steel, a line measuring 16 mm wide × 50 mm long × 0.75 mm deep can be hardened to 62 HRC in 0.25 seconds.

The process can be applied to most medium/high carbon or low alloy steels which are amenable to self-quenching. Successful laboratory trials have been conducted on such materials as AISI 01 tool steel, AISI 1045, 1075 and 1117, cast iron and high carbon sintered materials.

The high-frequency resistance heating technique would appear to offer strong competition to laser and electron beam methods of selective hardening. Like these two processes the new technique eliminates the need for external quenchants, thus minimising distortion of treated components. Unlike the electron beam technique it does not require a vacuum chamber, and there is no need for the work surface to be coated prior to treatment, as is frequently the situation with laser transformation hardening. Furthermore, the capital cost of the equipment is significantly less than that for laser or electron beam plant.

4. Flame hardening

Flame hardening, although displaced in many applications by induction hardening, is still widely used. The steels used and the degree of hardening obtained are similar to those in induction methods. Compared to induction hardening, flame hardening is more costly to operate and less easily controlled. However it requires less capital and there are few restrictions on the shape of components which can be treated.

The three main processes are:

(a) manual or spot;
(b) progressive;
and (c) spinning.

In the manual or spot method the flame is directed at the area to be heated until the required temperature is obtained and the area then quenched by immersion or in a spray. In the progressive method the flame is moved slowly over the surface and is followed by a quenching spray that may be attached to the burner head. The method is suitable for the hardening of large plates and long components such as lathe guides. In the spinning method the component to be hardened is rotated in the flame which is usually produced by

a number of burners. Like induction hardening the method is only applicable to axisymmetric or near axisymmetric parts such as gears and camshafts.

Fuels used include acetylene, propane and natural gas and these can be burned in air or oxygen. Of these acetylene has about twice and three times the combustion intensity (the product of the combustion velocity and the calorific value) of propane and natural gas respectively. Fuels are therefore selected on the basis of cost and the required speed and depth of hardening. If depths of hardening of less than 3 mm are required then fuel-oxygen mixtures are necessary, the rate of heat transfer from fuel-air mixture being too low. When using fuel-oxygen gas mixtures care must be taken to ensure that overheating resulting in excessive grain growth and oxidation does not occur.

5. Tungsten inert gas (TIG) hardening

In the proprietary Elowig system[4] TIG welding is used to harden cast iron by surface melting. This is an inexpensive method for the local hardening of cast iron and has been used on ferritic and pearlitic grey iron with flake graphite, nodular iron, malleable iron

Fig. 4.2 Comparison of costs of camshafts as foundry chill cast, induction hardened and ELOWIG hardened.[4]

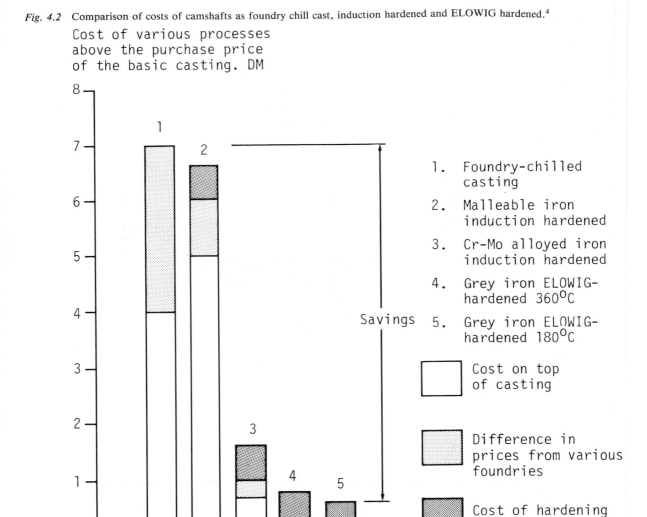

Cost of various processes above the purchase price of the basic casting. DM

1. Foundry-chilled casting
2. Malleable iron induction hardened
3. Cr-Mo alloyed iron induction hardened
4. Grey iron ELOWIG-hardened 360°C
5. Grey iron ELOWIG-hardened 180°C

Savings

Cost on top of casting

Difference in prices from various foundries

Cost of hardening

and chromium-molybdenum alloyed iron. Automobile camshafts treated by this method perform better than locally induction hardened shafts and compare favourably with chilled castings. Chilled castings are, however, about three times the price of grey iron-castings, whereas the cost of the Elowig process is only a relatively small premium on the grey iron price. This is illustrated in Fig. 4.2, where the costs, in Deutschmarks, of the various processes are shown as added costs to the basic cost of a grey iron camshaft. Castings are normally locally preheated in the range 200–400°C before treatment to increase the depth of melting and reduce the incidence of cracking.

6. Laser transformation hardening

Solid state transformation hardening methods using electroheat direct impingement (ie lasers and electron beams) are characterised by the fact that the surface of a ferrous material achieves the austenitisation temperature after a very short interaction time, whilst the bulk of the material remains unaffected. Consequently, when the heat source is removed, there is no need for an external quenchant to bring about martensite transformation to form the case, the bulk

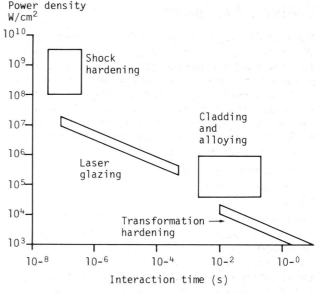

Fig. 4.4 Operational regimes for the laser heat treating of engineering materials.[5]

of the material providing an adequate heat sink to cause self-quenching. These factors are illustrated in

Fig. 4.3 Isothermal transformation diagram for a 0.4%C steel with a (superimposed) calculated temperature cycle at *a* surface and *b* depth 0.5 mm for a 10 mm thick steel sample subject to a power density of 3×10^3 W/cm² for a dwell time of 0.2 s.[5]

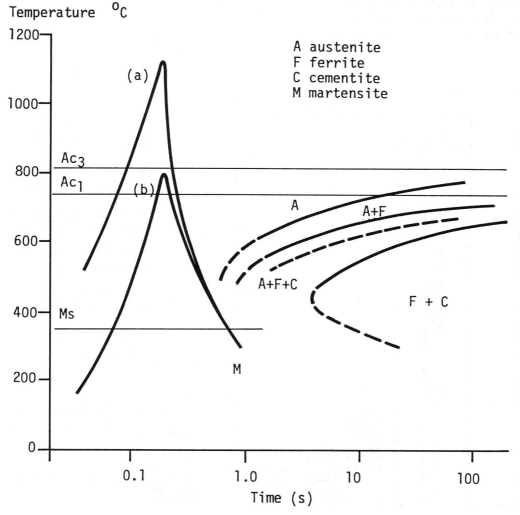

Fig. 4.3, for a 0.4%C steel. The power range used for solid state transformation hardening, as compared with other laser modification processes, is shown in Fig. 4.4.

The CO_2 laser which operates at a wavelength of 0.16 μm is the one most widely used in the heat treatment industry. Clean metallic surfaces are poor absorbers of radiation of this wavelength, as much as 95% of the incident energy being reflected and lost. It is normal practice, therefore, to cover a surface which is to be heat treated by lasers with a coating of, for example, colloidal graphite which is absorbing at the laser wavelength.

Typically, laser hardening is employed where cases up to about 1 mm are required. Much deeper cases can be achieved by reducing beam power density and treatment rates, but in this regime the process parameters are comparable with conventional low power density heating sources and the advantages of using a laser are less obvious.

Alloy composition, prehardened microstructure, and section thickness determine the hardness value and case depth for a given set of beam conditions. In general, all materials which transformation harden and are amenable to surface hardening by conventional techniques are suitable candidates for laser hardening. Since hardening cycles are extremely short, alloy irons and steels containing elements which enhance hardenability and hence reduce the critical cooling rate from the austenitising temperature, and ferrous materials which readily austenitise are those materials best suited to laser heat treatment. However,

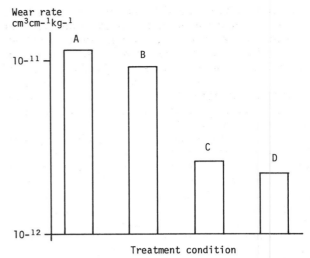

A. As-cast (pearlitic)
B. Retained austenite + coarse martensite
C. Fully martensitic
D. Surface melted (ledeburitic)

Fig. 4.6 Comparison of abrasive-wear rates of microstructures arising from various laser-hardening treatments of grey cast iron.[6]

because the laser process is very localised and can result in extremely rapid cooling rates, even low carbon steels (0.2%C) can be transformation hardened.

The microstructure of a cast iron after treatment, shown in Fig. 4.5,[5] and the accompanying microhard-

Fig. 4.5 *a* Micrograph showing the laser hardened case formed on cast iron; IPD 2.5×10^3 W/cm², coverage rate 160 cm²/min, etched in picral; *b* microhardness profile after treatment.[5]

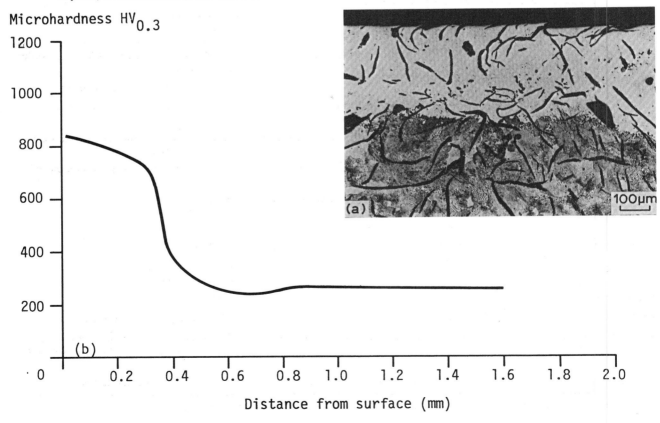

ness profile reveal that the transformation hardened layer is fully martensitic. Liquid phase transformation hardening, or laser glazing, is also possible with Grade 17 cast iron. In this treatment a finely dispersed ledeburitic structure, with a hardness around 1100 HV, is formed on the outermost surface with a martensitic sub-zone. The abrasive wear rates of the various structures formed on cast iron by laser treatment are compared in Fig. 4.6.[6]

In 1981 a study of the cost of operating CO_2 lasers for selective surface hardening was undertaken by IIT Research Institute.[7] It was concluded that operating costs ranged from $43 to $250 per 100 in², the lower costs requiring fully automated tooling, careful control of changeover time (in a contract heat treatment shop) and close component inspection.

Applications of laser surface hardening include ferritic malleable steering gear housings, AISI 1050 motor shaft splines, cutting blades in 1050 steel, track parts for tanks, typewriter interposer bars and electric razor cutter combs in 0.7%C steel. Laser hardening of steering gear housings has been discussed in Chapter 3.

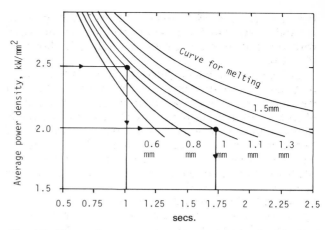

Fig. 4.7 Electron beam treatment parameters to harden a 0.38%C steel to 500 HV at the depths shown.[9]
(This assumes a static workpiece with respect to the beam, of about 1 mm diameter at the point of impact, which scans at 100 points/cm² to produce a uniform power density. For example, treatment of a surface area of 3 cm² to 500 HV at a depth of 1 mm requires a 6 kW beam (3×2 kW) operating for 1.7 seconds or a 7.5 kW beam (3×2.5 kW) operating for 1.05 seconds).

7. Electron beam transformation hardening

Electron beam technology has been available for about twenty-five years and is now widely accepted as a reliable industrial tool, especially for welding.

The advent of laser transformation hardening techniques, with the attendant fine focused high energy beam, has led to interest in applying electron beams for the same purpose. The main advantage of laser treatment is the ability to operate in air without the need for vacuum, which is essential for the electron beam technique. On the other hand, vacuum provides an excellent environment to protect the heat treated

Table 4.3   Comparative aspects of laser and electron beam hardening as seen by a user[8]

Item	Electron beam	Laser
Bore size limitation	Line of sight at 35° impingement	1 in dia. and up—no limit. ¼ in to 1 in—using small spot size.
Part contamination	Parts should be clean	Parts should be clean at reaction area
Focused spot size	Hard vacuum—0.020 in dia. Soft —0.030 in dia.	Varied—depends on system and optics
*Heat treat pattern	Shape and density controllable by computer or function generator	Shape limited to mirror deflection capabilities. Density averaged.
Beam deflection	Electromagnetically 30° included—two axes	Water cooled reflecting optics—mechanical or integrating
Effect of vacuum on cycle time	Production systems pre-pumped—no effect. Low volume systems—minimum 5 sec pump down, but depends upon chamber size.	No effect
Part size limitations	Limited to chamber size and motion required	No limit
Power available	100 kW +	15 kW max (commercial) 100 kW + (experimental)
Conversion efficiency	>90%	<10%
Operating costs	Low	High due to gases and low efficiency
Surface preparation	None—can be done on finish ground surfaces	Can be done on finish ground surfaces but requires an absorptive coating that may need cleaning
Investment	Lowest on high power applications	Lowest on low power applications

*Pattern size is controlled by the available power of the equipment. Nominally, 10–20 kW/in² is used.

surface. Features of laser and electron beam surface hardening techniques are compared in Table 4.3.[8] The capital and operating costs of hardening with electron beams are much lower than with lasers.

By 1981, nine electron beam machines had been installed in US automobile plants. In the best documented of these,[9] automatic transmission clutch cams, in SAE 5060 steel, are selectively hardened to a depth of 1.5 mm at a rate of 250 per hour. The process was introduced as the answer to distortion problems experienced with induction hardening.

Other aspects of electron beam heat treatment which are discussed[9] include:

(a) the depth of the heat treated layer can be from 0.2 mm to 2.5 mm, with best results in the range 0.75 to 1 mm. For self-quenching the thickness of the part must be at least four times that of the heat treated layer in order to provide a sufficient heat sink.

(b) the hardness which can be achieved by electron beam treatment can be 2–3 points HRC higher than that achieved by induction hardening.

(c) the data shown in Fig. 4.7, can be used, as a first approximation, to define the power of the beam needed to achieve a hardness of 500 HV at the bottom of a given layer in a 0.30%C steel.

(d) with regard to the relative position of beam and component, the impingement of the beam can be accurate to ± 0.05 mm.

The applications of electron beam surface hardening are not yet extensive but, in principle, many of the components being investigated for laser treatment are potential candidates.

References

[1] GUTHRIE, A M and ARCHER K L. Induction surface hardening. *Heat Treatment of Metals*, Vol 2, (1), p 15, 1975

[2] DAVIES, J and SIMPSON, P. *Induction Heating Handbook.* McGraw Hill, NY, 1979.

[3] HICK, A J. Selective surface hardening. *Heat Treatment of Metals*. Vol 9, (1), p 10, 1982.

[4] REINKE, F H. TIG hardening of cast iron cam-shafts. *Heat Treatment of Metals*. Vol 8 (1), p 17, 1981.

[5] TRAFFORD, D N H, BELL, T, MEGAW, J H P C and BRANSDEN, A S. Heat Treatment using a high power laser. *Heat Treatment '79*. p 32, The Metals Society, London, 1980.

[6] TRAFFORD, D N H, BELL, T, MEGAW, J H P C and BRANSDEN, A S. Laser treatment of grey iron. *Heat Treatment '81*, p 198. The Metals Society, London, 1983.

[7] SEAMAN, F. Cost considerations in laser heat treatment. *6th ASM Heat Treating Conference/Workshop*, Sept 1981.

[8] HICK, A J, Rapid surface heat treatments—a review of laser and electron beam hardening. *Heat Treatment of Metals*, Vol 10, (1), 1. p 3, 1983.

[9] SAYEGH, G. Principles and applications of electron beam heat treatment. *Heat Treatment of Metals*, Vol 7 (1), 1, p 5, 1980.

Additional reading

THELNING, K E. *Steel and its heat treatment.* Butterworths, London, 1984.

Metals Handbook, 9th edn., Vol 4. —Heat Treating. ASM publication 1980.

CHAPTER 5
Thermochemical treatments

1. Introduction

Thermochemical treatments are those in which elements, principally carbon and nitrogen, are introduced into the surface at elevated temperatures using solid, liquid, gas or plasma transfer media. Processes include austenitic carburising and carbonitriding methods which, like thermal treatments, rely on the surface transformation to martensite to produce a wear resistant surface; and ferritic thermochemical treatments, nitriding and nitrocarburising, which are carried out at temperaturers below the formation of the austenite phase and harden by means of a fine scale dispersion of alloy nitrides and carbonitrides in the matrix. Most thermochemical processes are applied to steels and cast irons but some are applicable to non-ferrous materials. The principal features of the main austenitic and ferritic thermochemical treatments are summarised in Table 5.1.

Gaseous thermochemical treatments are by far the most widely used, particularly in high volume continuous production. They are generally cheaper to operate than liquid salt baths which can present environmental and component cleaning problems. On the other hand, the atmospheres used for gaseous treatments are frequently potentially explosive so that appropriate safety measures must be observed. These include the use of nitrogen-based controlled atmospheres, or automatic nitrogen purge systems. In addition, equipment for gaseous treatments is available in larger capacity units than salt baths, and the process provides more versatility in that a change from one process to another may involve only a change in gas composition and operating temperature.

Plasma thermochemical processing, in particular plasma nitriding, has grown in importance over the last ten years and it is estimated that more than 500 units are now in production on a world-wide basis.

Other thermochemical processes involve the diffusion of boron into metal surfaces and of metals such as niobium and vanadium into steels to form very hard carbide layers on the surface.

2. Carburising

The carburising process involves heating steel components to a temperature, usually in the range 850–950°C, which is above the ferrite/austenite transformation temperature in a solid, liquid or gaseous carburising medium so that the surface becomes enriched with carbon. The parts are then quenched to

Table 5.1 Principal features of thermochemical treatments involving the diffusion of carbon and nitrogen

Process	Description	Process	Typical treatment temp, °C	Typical case depth, mm	Typical surface hardness, HV
Carburising	A process in which a steel surface is enriched with carbon, at a temperature above the ferrite/austenite transformation. On subsequent quenching, an essentially martensitic case is formed	Solid, Liquid Gaseous Plasma	850–950	0.25–4.0	700–900*
Carbonitriding	Similar to carburising, but involving nitrogen as well as carbon enrichment	Liquid Gaseous Plasma	750–900	0.05–0.75	600–850*
Nitrocarburising	A process in which a steel or cast iron surface is enriched with nitrogen, carbon and possibly sulphur at a temperature below the ferrite/austenite transformation	Liquid Gaseous Plasma	570	0.02 max[†] 1.0 max[‡]	500–650[†]
Nitriding	A process in which a steel surface is enriched with nitrogen, at a temperature below the ferrite/austenite transformation	Gaseous Liquid Plasma	500–525	0.4–0.6	800–1050

*Depending on tempering treatment (upper figure represents typical as-quenched hardness).
†Thickness and microhardness of compound layer on mild steel. Values are dependent on alloy content of material treated.
‡Total depth of diffusion zone.

develop an essentially martensitic structure in the surface layer or case, the hardness of which increases with carbon content up to the eutectoid level (0.8%C in plain carbon steels) Fig. 5.1. Meanwhile the core material which has a lower hardenability because of its lower carbon content transforms to a softer more ductile structure, eg ferrite and carbide, or possibly a low carbon bainite or martensite depending on the hardenability of the steel. Normally the hardened parts are tempered at 150–200°C. As a result of the compressive stresses developed in the outer surface layers during carburising the fatigue strength is also increased substantially.

Depending on the steel composition and the process used, the carburised surface may have a hardness up to 900 HV and a depth of up to 4 mm. Case depth may be expressed either as total or effective depth of carbon penetration. The total depth is usually determined by chemical analysis or metallography, the effective depth is that above a specified carbon content or hardness level, usually 550 HV. With the majority of carburised components the choice of case depth is a matter of experience, accurate calculation being difficult because of the complex nature of the loads exerted on the surface. In general terms the case depth is insufficient if the maximum shear stress as a result of the applied load occurs in the low carbon region, ie at or below the case-core interface. Under such circumstances cracking will occur in this region. The remedy is either to increase the case depth or increase the core strength (eg increase the hardenability of the steel used). The strength of steel at the carburising temperature is comparatively low and to avoid deformation by creep, it may be necessary to provide structural support using appropriate fixtures. Distortion may also occur as a result of structural changes in the case and/or thermal stresses in the component. The nature and magnitude of the dimensional changes depend, on the depth of the case, the method of hardening and the configuration of the component. To minimise expensive post treatment grinding it is usually necessary to carry out preliminary tests to establish the magnitude of the dimensional change so that allowances can be made in component design.

Because carburising is applicable to both plain carbon and alloy steels giving a wide range of core strengths and is the only thermochemical treatment which can provide very deep cases it is utilised in diverse applications. It has been estimated that carburising accounts for between 70–80% of the components treated thermochemically. It is used extensively on components for the automobile industry where a high surface hardness on easily cold formed or machined low alloy steels is required, for heavily loaded components such as gears, and in roller bearing races which require high contact fatigue resistance.

(i) Carburising steels

Of the usual alloying elements, chromium and manganese are subject to oxidation in endothermic gas atmospheres whereas molybdenum and nickel are not. Consequently in gas carburising selective oxidation of

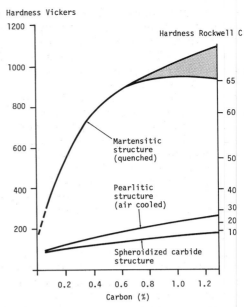

Fig. 5.1 Variation in hardness of the microconstituents of steel with carbon content.

chromium and manganese may reduce alloy content at the surface with a lowering of hardenability. Chromium and manganese also reduce the carbon content of the eutectoid and increase the tendency to form hypereutectoid carbides in the case. For these reasons chromium and manganese are seldom over 1% in carburising steels, increased hardenability being obtained by nickel and molybdenum additions.

The hardening treatment employed may consist of:

(a) quenching from the carburising treatment (direct quenching);

(b) slow cooling from the carburising temperature to room temperature, followed by reheating (usually to 780–820°C) and quenching (single quenching);

(c) direct quenching, followed by re-heating to a lower temperature (usually 780–820°C) and quenching (double quenching).

Direct or single oil quenching is normally used. Double quenching is used with some alloy steels to produce a better case hardness profile in deep cases.

Core strengths of carburised steels can range from 430 to 1310 MN/m² (28 to 85 tsi) in the hardened and tempered condition. The choice of steel composition depends largely on the properties required in the core, more highly alloyed types being required in highly stressed applications. Some examples of case hardening steels and their applications are given in Table (5.2).[1] Where high surface hardness alone is required, a plain carbon steel is usually adequate but to achieve full case hardening it may be necessary to water quench from the hardening temperature, a procedure which may give rise to distortion or cracking, especially in components of non-uniform cross-section.

(ii) Carburising processes

(a) Pack carburising

In pack carburising the components to be treated are

put into steel boxes along with a carburising agent such as coke and an energiser, eg barium carbonate. The boxes are then heated in a furnace to a temperature in the range 850°C to 950°C. Control of the treatment is possible, in principle, by adjusting the amounts of carburising agent and energiser employed. However, in practice this can be difficult to achieve and may result in poor mechanical properties after quenching, due to incorrect surface carbon levels. Quenching directly from the carburising temperature is not possible, thus making re-austenitising and quenching normal practice. In energy terms this additional operation makes the process inefficient but nevertheless the process can be used to advantage on a limited scale for 'one-off' components.

(*b*) *Salt bath carburising*

Engineering components can be carburised in molten salt baths containing mixtures of sodium cyanide and an alkaline earth salt, eg barium chloride. The probable reaction sequence is the oxidation of the cyanide to cyanate which subsequently dissociates at the steel surface

$$4Na\ CNO = 2NaCN + Na_2CO_3 + CO + 2N_{Fe}$$

The carbon monoxide, in turn, reacts with the steel

$$2CO = CO_2 + C_{Fe}$$

(where the suffix $_{Fe}$ denotes solution in iron)

As the operating temperature and/or the cyanide content is raised, the carburising potential increases and the nitrogen potential decreases (Fig. 5.2 & 5.3).[2] Salt baths are normally operated around 850°C and are usually used for producing cases less than 0.5 mm

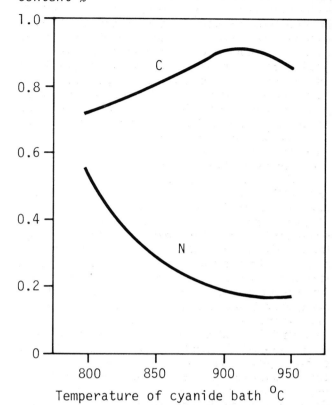

Fig. 5.2 Effect of carburising temperature on surface composition of plain carbon steel treated for 2.5 hrs. in liquid bath containing 50% NaCN.[2]

Table 5.2 Examples of case hardening steels and their applications[41]

Specification (BS970 1971/2	Typical Composition(%)					Characteristics and Uses
	C	Mn	Ni	Cr	Mo	
070M20 (En3)	0.20	0.70	—	—	—	Machine parts requiring hard wear resistant surface with tough core— 35TSI UTS in ¾″ LRS.
665M17 (En34)	0.18	0.55	1.75	—	0.25	For applications requiring combined hardness and toughness eg small gears and pins. 50TSI UTS in ¾″ LRS.
655M13 (En36A)	0.13	0.48	3.25	0.85	—	For heavy load applications requiring core strength/toughness and high surface hardness eg high duty gears, worm gears, crown wheels etc, 65TSI UTS in ¾″ LRS.
659M15 (En39A)	0.15	0.38	4.00	1.20	—	Used for extremely demanding applications or where ruling sections are large, eg crown wheels, large traction gears etc, 85 TSI UTS in ¾″ LRS.
530H30 (En18A)	0.30	0.70	—	1.00	—	Used in small sections where core toughness is less important than high surface hardness & wear resistance. Alternatively may be used in large sections to provide deep effective case depths more economically.

LRS—limiting ruling section.

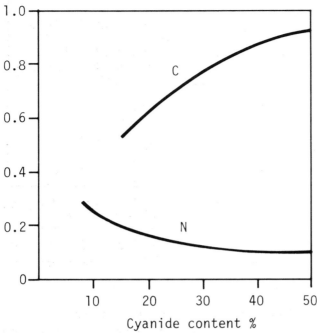

C and N content %

Cyanide content %

Fig. 5.3 Effect of cyanide content on surface composition of plain carbon steel treated at 950°C for 2.5 hrs. in liquid bath containing 50% NaCN.[2]

thick. By increasing the bath temperature to 950°C case depths of up to 2 mm can, however, be produced.

The rapid heat transfer characteristics of the salt bath make it ideal for treating large numbers of small components. However, problems arise over the cleaning of complex components and the disposal of residual toxic waste materials. For these reasons salt bath processes have been largely superseded by gaseous processes.

(c) Gas carburising

Various gaseous carburising techniques account for the bulk of production carburising carried out. This is largely due to the ease with which the surface carbon level can be controlled when compared with pack and salt bath methods. Although carburising atmospheres may be generated in many ways, the basic furnace reactions remain unaltered. It is normal practice to generate, either inside or outside the furnace, a carrier or background gas of low carbon potential. Enriching additions of hydrocarbons are then made to increase the carbon potential to the required level. The carbon potential can be defined as the surface carbon level achieved on a steel in a particular atmosphere under equilibrium conditions.

ENDOTHERMIC GAS ATMOSPHERES

In this treatment the carrier gas is generated outside the furnace. Methane or high purity liquid propane is mixed with air and burnt in a controlled manner in a gas generator producing a gas typically of the following composition:

N_2	H_2	CO	CO_2	CH_4
35–40%	40–45%	15–25%	0.1–1.0%	0.5–1.5%

Controlled additions of methane or propane are then made to increase the carburising potential of the atmosphere.

$$CH_4 + CO_2 = 2CO + 2H_2$$
$$CH_4 + H_2O = CO + 3H_2$$

At the carburising temperature (usually 880–930°C) several carburising reactions take place of which the most important is

$$CO + H_2 = C_{Fe} + H_2O$$

The balance between the atmosphere's constituents is maintained by the water gas reaction

$$CO + H_2O = CO_2 + H_2$$

which is the fastest of all the furnace reactions.

In an alternative process involving the same basic principles, the carburising atmosphere is produced by dripping liquid hydrocarbons, eg benzene and oxygenated hydrocarbons, eg alcohols, onto a hot plate within the furnace.[4] Sooting tends to be more of a problem in this process because of inefficient cracking of the reactants and forced circulation by fans is essential to ensure that the atmosphere is evenly distributed throughout the furnace.

NITROGEN BASED ATMOSPHERES

With the cost of fossil fuels increasing in the 1970s interest developed in nitrogen-based atmospheres for carburising, the aim being to reduce the proportion of combustible gaseous constituents to a minimum and to eliminate the need for gas generators.

The carburising atmospheres are generated by passing a mixture of nitrogen, hydrogen, carbon and oxygen-bearing gases through a mixing panel into the reaction chamber. Carbon dioxide and air are typically used as oxygen-bearing gases and oxidise the hydrocarbon, typically methane, to carbon monoxide and hydrogen. Control of the carbon potential is achieved by regulating the supply of hydrocarbon to the system. In contrast to endothermically generated atmospheres these atmospheres may contain 80% nitrogen and only 20% combustibles.

ATMOSPHERE CONTROL

Earlier time-consuming methods such as complete gas analysis and the use of shim stock for determining carbon potential have now been displaced by instrumental methods which give continuous information on the state of the furnace atmosphere and allow any necessary adjustments to be carried out. These methods are based on the fact that the furnace atmosphere is controlled by the reaction,

$$CO + H_2O = CO_2 + H_2$$

and that by measuring the water content (by dew point) or the CO_2 content (by infra-red) or the oxygen potential (by a zirconia probe) the CO content and thus the carburising potential of the gas can be determined. Although carburising is essentially a non-equilibrium process these methods have proved to be very useful in atmosphere control.

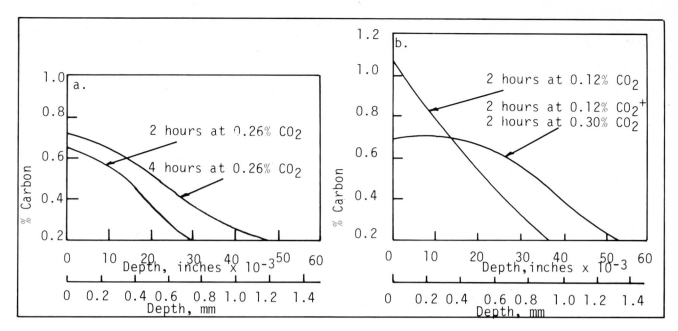

Fig. 5.4   Carbon profiles resulting from single stage and boost-diffuse carburising treatments at 950°C.[3]

BOOST-DIFFUSE

Boost-diffusion treatments are used to produce the desired surface carbon content and effective case depth in the shortest possible time. In the initial boost stage of the process the carburising potential of the atmosphere is raised to give a surface carbon content above that required to give optimum surface hardness. In the subsequent diffusion stage, the surface carbon level is reduced to that required for optimum mechanical properties and the case depth increased at the same time. The carbon profiles resulting from a single stage and a boost-diffuse treatment on similar steels are compared in Fig. 5.4.[3] The use of this technique to produce carburised cases greater than 0.5 mm can result in time savings of up to 70% for alloy and 30% for plain carbon steels.

(iii) Process developments

(a) Fluidised bed carburising

Controlled atmosphere fluidised beds, with their high heat transfer rates, have been explored as a possible alternative to salt baths for carburising.[5] Fuel consumption is lower than that of a salt bath of similar capacity and the low thermal capacity of the bed means that it can be used economically for intermittent operation. The non-wetting characteristics of the bed eliminate 'drag-out' and the need for washing parts after treatment. The process is versatile and can be used to perform a neutral hardening operation followed by a carburising treatment, in successive cycles, by adjustment of the atmosphere's composition.

Rates of carburising are about 30% higher than in conventional gas carburising but accurate control of the atmospheres and therefore of the carbon potential has proved difficult and for this reason the process has only found limited application.

(b) Low pressure (vacuum) carburising

In this process components are heated to the carburising temperature under a pressure of 0.1 torr. When the carburising temperature has been reached the furnace is backfilled with a hydrocarbon gas such as methane or propane to the desired partial pressure. The rate of carburising as a function of methane partial pressure is illustrated in Fig. 5.5[6] Although

Fig. 5.5   Effect of methane partial pressure on carbon profiles obtained by vacuum carburising at 950°C.[6]

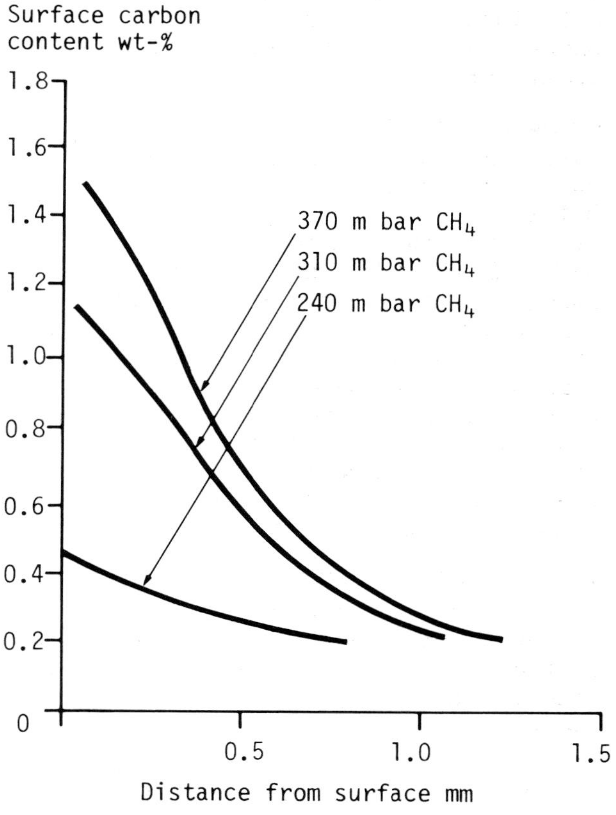

Surface carbon content wt-%

some process control can be achieved by adjusting the methane partial pressure, soot formation and the lack of uniformity of the carburised case have severely limited the industrial acceptance of this process.

(c) Plasma carburising

In plasma carburising the workpiece is raised to the treatment temperature by resistance heating in a low pressure argon atmosphere.[7,8] A hydrocarbon (methane or propane) plus hydrogen gas mixture is then introduced to give a chamber pressure in the range 1–20 m bar. Application of an electrical potential of several hundred volts between the cathode (components) and the anode (chamber) ionises the gas and produces a glow discharge which activates the carburising atmosphere and causes very rapid carbon mass transfer to occur. This is illustrated in Fig. 5.6[9] where the carburising effect of the atmosphere with and without the discharge is shown.

Materials carburised by this technique exhibit very steep carbon concentration profiles even at short treatment times (Fig. 5.7)[7] which allows boost/diffuse principles to be employed with consequent reduction in treatment times. Uniform case depths are obtained on components of complex geometry.

Control of the process is achieved by adjusting the composition of the treatment atmosphere, the electrical characteristics of the glow discharge and the treatment temperature which can vary between 800 and 1100°C.

Plasma carburising has all the advantages of vacuum heat-treatment without many of the limitations of the vacuum carburising process. Some prototype units with batch furnaces and integrated sealed quench facilities are currently in use and larger industrial units are planned.

3. Carbonitriding

Carbonitriding is a variation of the carburising process in which up to 0.5% nitrogen as well as carbon is introduced to the steel surface thereby increasing hardenability and reducing the austenite-martensite transformation temperature. The increased hardenability enables a case of more uniform hardness to be obtained on plain carbon and very low alloy steels than is produced by carburising. A less severe quenchant can sometimes be used. In steels of higher hardenability, however, the nitrogen content must be controlled to ensure that the amount of retained austenite in the case is within acceptable limits.

The carbonitriding process is usually applied to plain carbon and very low alloy steels to produce relatively shallow cases (<0.75 mm). In these steels the core is not significantly strengthened during heat treatment so that carbonitrided components are not suitable for high load applications. Apart from this point the process results in property improvements very similar to carburising and is used extensively in the automotive industry, especially on shafts and transmission parts, where high surface hardnesses and enhanced wear resistance are required. In addition, very large volumes of metal stampings in low cost mild

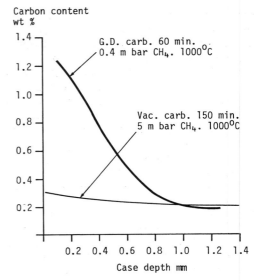

Fig. 5.6 Comparison of carbon profiles with and without glow discharge (G.D.) during carburising.[9]

steel are regularly subjected to a high temperature carbonitride treatment.

Because the processing temperature range, 750°C–900°C, is lower than that for carburising, the resultant shape distortion can be less and hence carbonitriding is frequently specified for low distortion applications. However, with pressed parts stress relieving during treatment can still lead to excessive distortion and hence there is a trend to replace, at least in part, the austenitic carbonitriding treatment, by the very low distortion ferritic nitrocarburising.

(i) Carbonitriding processes

(a) Salt bath carbonitriding

This process is similar to liquid carburising but the amount of oxygen supplied to the bath is increased which increases the amount of sodium cyanate in the bath which in turn leads to greater production of nitrogen.[10] The process is also carried out at a lower temperature than carburising which leads to a proportionately greater increase in nitrogen content (Fig. 5.2).

The degree of control over the process is not particularly good owing to the complex nature of the salt bath. The efficiency of the bath can only be maintained by regular analysis and adjustment of composition. Despite these problems the process finds regular use in industry. It is cheaper than liquid carburising owing to the lower operating temperature range.

A proprietary process, Noskuff, improves the frictional and anti-scuffing properties of steels which contain at least 0.6%C or which are pre-carburised.[11] The process consists of a 10–15 minute treatment usually at 730–750°C, in a fused salt bath containing cyanide and at least 10% sodium cyanate, when a nitrogen rich compound layer and an underlying nitrogen rich diffusion zone is formed. On subsequent quenching a considerable amount of austenite is retained beneath a hard surface layer rich in carbides and nitrides.

(b) Gas carbonitriding

In this process ammonia gas is introduced into the carburising atmosphere to increase the nitrogen potential. Typically the process is carried out at 850°C in sealed-quench furnaces using an ammonia concentration around 5%.[12] At higher temperatures the nitrogen potential of ammonia additions is reduced. This is illustrated in Fig. 5.8 together with the effect of ammonia additions on the carbon potential of a carburising gas atmosphere.

Control of the ammonia concentration is important in that high levels can lead to retained austenite and/or void formation in the case.[13] Nitrogen, in addition to increasing the hardenability of steels lowers the austenite-martensite transformation temperature and high levels, using normal quenching procedures, can lead to retained austenite. Sub-surface void formation is thought to be due to the formation of molecular nitrogen at grain boundaries and aluminium killed steels in the cold worked condition are particularly susceptible.

4. Nitriding

Nitriding is a thermochemical treatment in which nitrogen is diffused into a steel surface, at a temperature usually within the range 490–530°C. (At lower temperatures rates of nitriding are impractically slow and at higher temperatures there is a reduction in hardness due to precipitate coarsening and also tempering of the core.) Atomic nitrogen, formed by the catalytic dissociation of ammonia at the steel surface or in a nitrogen-hydrogen plasma diffuses into the steel with the formation of a superficial white layer consisting of a mixture of a face centred cubic γ' phase (based on Fe_4N) and a close packed hexagonal ϵ phase of a higher nitrogen content (Fig. 5.9).[14] As the thickness of this layer builds up, nitrogen diffuses from it into the underlying steel to form finely dispersed needles of alloy nitrides and, by reaction with carbon in the steel, carbonitrides, some of which may form at grain boundaries (Fig. 5.10).[15] The white layer is thin (~0.02 mm) but brittle and, to prevent spalling in service, it is usually removed by grinding or by chemical treatment, both of which are relatively expensive. The underlying case which is usually 0.2–0.7 mm thick has a peak hardness of 900–1100 HV, excellent abrasion resistance, and gives rise to a marked improvement in fatigue performance. The case is temper-resistant up to ~500°C. No phase transformation is involved in cooling to room temperature, and quenching is not required to develop surface hardness. Owing to the relatively low processing temperatures employed, there is less distortion than in carburising or carbonitriding. However, the surface compressive stresses developed in the case on nitriding, which improve fatigue resistance, can also lead to distortion in hollow thin-walled components. Consequently, nitriding is best carried out on substantially rigid, preferably axi-symmetrical components such as shafts, thick walled cylinders, gears and die blocks. The process has been widely applied to improve the wear resistance of barrels and extruder screws in plastics machinery, and of diesel engine components such as crankshafts, valve stems and injectors.

Fig. 5.7 Carbon profiles obtained in various materials by plasma carburising.[7]

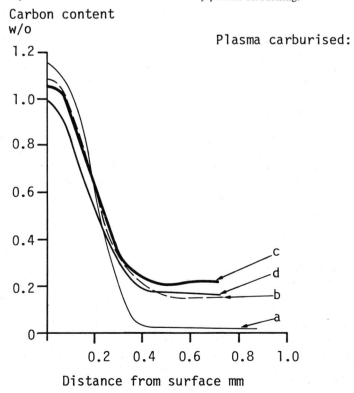

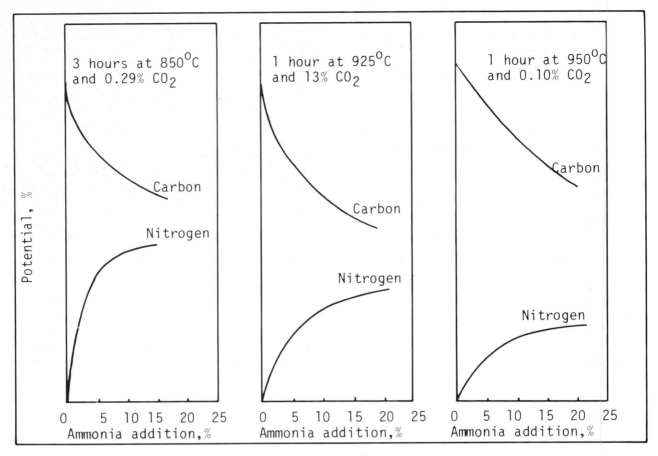

Fig. 5.8 Effect of ammonia additions on nitrogen and carbon potentials.[13]

(i) Nitriding steels

To achieve an appreciable hardening effect on nitriding, steels must contain nitride-forming elements such as aluminium, titanium, chromium, molybdenum and vanadium.[16] Aluminium and titanium have a pronounced effect on hardness, even at small concentrations ($\sim 1\%$). Chromium, in small concentrations, molybdenum and vanadium have a lesser effect. As the concentration of chromium increases so does its hardening capacity. Increasing the concentration, especially of the strong nitride-formers, results in a decrease in the effective case depth. Examples of steels commonly used for nitriding are given in Table 5.3, and typical hardness depth curves in Fig. 5.11.[17]

Steels must be hardened and tempered before nitriding to produce the desired core properties and a structure that is stable at the nitriding temperature. The tempering temperature has a pronounced effect on the hardness and case depth of the nitrided steel. The effect of increasing the tempering temperature of En19 steel from 550°C to 700°C is to reduce the peak hardness from about 780 to 600 HV whilst the effective case depth (to 500 HV) is halved to 0.25 mm. Thus a degree of flexibility in the hardness profile is possible. The effect is illustrated in Fig. 5.12.[18]

(ii) Nitriding processes

(a) Gas nitriding

Gas nitriding is usually carried out in batch type furnaces with the charge contained in gas-tight nickel alloy retorts through which pure anhydrous ammonia is circulated. Processing temperatures are in the range 490–530°C and are maintained within $\pm 5°C$ to ensure reproducibility. Part of the ammonia dissociates catalytically on the surface of the components and some of the nascent nitrogen is absorbed by the steel

$$2NH_3 = 2N_{Fe} + 3H_2$$

A typical nitriding cycle involves 45 h at a nitriding temperature of 515°C with a total cycle time of 56 h. Components after undergoing the above nitriding cycle would have a case depth of about 0.5 mm and be a matt grey colour due to the compound layer which has formed on the surface and which on components such as crankshafts subjected to high loads is removed by grinding and lapping or chemical techniques using cyanide based solutions. Increased labour costs and environmental concern has led to a re-appraisal of methods of minimising the amount of 'white-layer' formed. The most effective of these is reduction of the nitriding potential defined as

$$r = \frac{P_{NH_3}}{(P_{H_2})^{3/2}}$$

by diluting the ammonia with hydrogen. A typical threshold nitriding potential curve is shown in Fig. 5.13. However very close control of the nitriding potential is necessary if the hardness profile is not to be significantly reduced[15] (Fig. 5.14).

(b) Plasma Nitriding

The acceptance of plasma nitriding on an industrial scale lies, at least in part, on the contribution the process can make towards accurate control of the compound layer quality and thickness. Plasma nitriding, which has alternatively been called 'ion nitriding'[19] or 'glow discharge'[20] nitriding in the technical literature, is now widely used as a technically and, in many situations, economically viable alternative to gas nitriding by a variety of manufacturing industries. It can be used to surface harden most grades of tool steels, stainless steels, cast irons, many low alloy steels and titanium and its alloys.[21]

Fig. 5.15[22] shows a schematic diagram of a plasma nitriding furnace. The components to be treated are placed in the reaction vessel and cathodically charged to a potential in the range 400–1500 V, with respect to the wall of the vessel, which acts as the anode of the circuit. At the working pressure of about 0.1–10 torr a glow discharge of the reaction gases, hydrogen and nitrogen, is created. Nearly all of the potential drop across the 'vacuum' occurs within a few millimetres from the surface of the workpiece and this region appears as a luminous glow fringe. This glow fringe follows the contours of the component and so ensures

Fig. 5.10   Metallographic section of nitrided steel (X100).[15]

uniform nitriding irrespective of its shape. By adjustment of gas pressure and electrical parameters, blind holes and long narrow bores can be nitrided. The bombardment of the components by the treatment gas ions provides the heat input for the process, no

Fig. 5.9   Iron-nitrogen equilibrium diagram.[14]

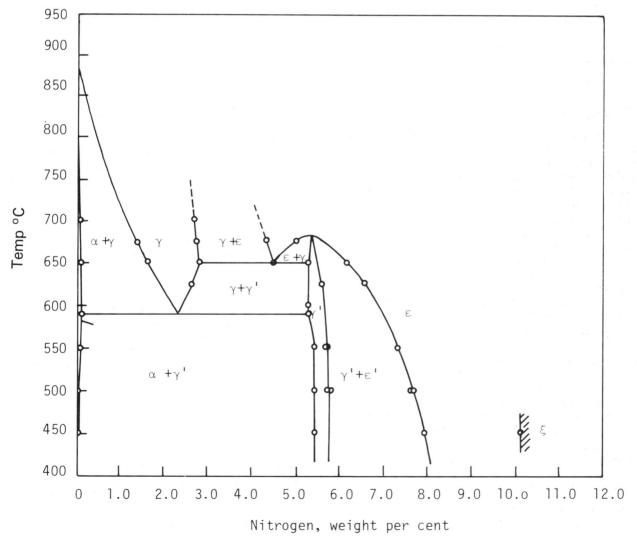

Table 5.3 Representative steels suitable for nitriding[17]

Steel type	BS 970 Grade	En	Chemical composition, %* C	Mn	Si	Cr	Mo	Al	V	Nearest equiv. SAE No.
Aluminium-chromium-molybdenum (Nitralloy)	905 M31	41A	0.27–0.35	0.40–0.65	0.10–0.45	1.4–1.8	0.15–0.25	0.9–1.3	—	7140
	905 M39	41B	0.35–0.43	0.40–0.65	0.10–0.45	1.4–1.8	0.15–0.25	0.9–1.3	—	
1% chromium-molybdenum	709 M40	19	0.36–0.44	0.70–1.00	0.10–0.35	0.90–1.20	0.25–0.35	—	—	4140
3¼% chromium-molybdenum-vanadium	897 M39	40 C	0.35–0.43	0.45–0.70	0.10–0.35	3.0–3.5	0.80–1.10	—	0.15–0.25	—
5% chromium hot work die steel (BH13)	—	—	0.32–0.42	0.4 max	0.85–1.15	4.75–5.25	1.25–1.75	—	0.9–1.10	H13

*as quoted in BS 970, Part 2, 1970 and BS 4659, 1971

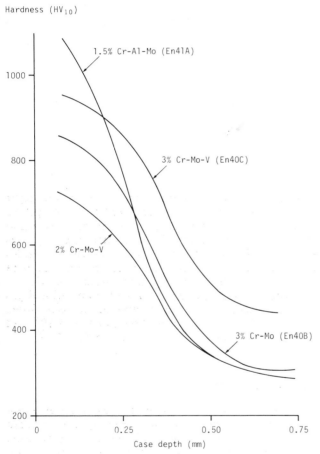

Hardness (HV$_{10}$)

Fig. 5.11 Hardness profiles of representative steels nitrided for 80 h at 500°C.[17]

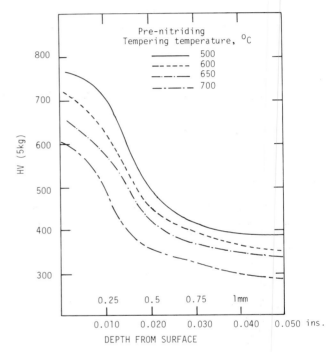

Fig. 5.12 Effect of pre-nitriding tempering temperature on hardness gradients in En19 steel.[18]

external heat source being required. Process control parameters are the composition of the gas mixture, pressure, voltage, current density and cathode temperature. This last is usually within the range 400–550°C and is controlled by the power input. As the nitriding potential is governed by electrical parameters, plasma nitriding can be carried out at lower temperatures than is the case with gas nitriding. Advantage can be taken of this in treating materials such as maraging steels when a temperature at which age hardening and nitriding take place simultaneously can be selected.

Some of the advantages of plasma nitriding over conventional gas nitriding are:
  —reduction in treatment gas consumption.
  —in most cases reduced energy consumption (Fig. 5.16).[23]
  —approximately 30% reduction in process time.
  —ability to remove oxide films from high chromium steels by sputtering prior to nitriding.
  —reduction in thickness of compound (white) layer to 6–8 μm (Fig. 5.17)[22] from 20 μm

In addition to reduction in the thickness of the white layer formed some control can be exercised over its composition by adjustments in gas composition, pressure and component temperature. Conventional gas nitriding results in a compound layer which consists of a mixture of γ′ and ε phases. It is possible that the friable nature of these layers is due to the stresses in the transition region between the different crystal structures. By controlling plasma nitriding par-

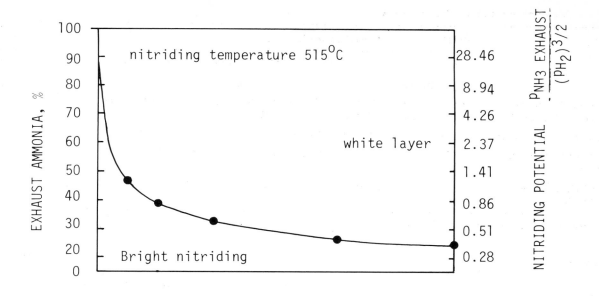

Fig. 5.13 Variation of white layer formation with nitriding potential.[15]

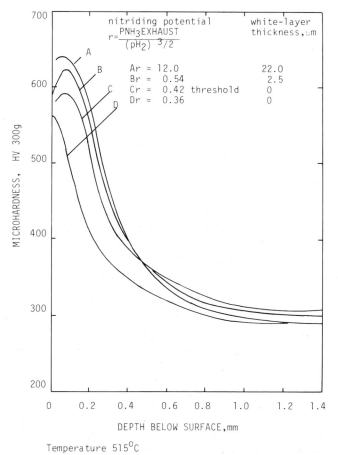

Fig. 5.14 Variation of hardness profile with nitriding potential.[15]

ameters an essentially monophase γ' compound layer of the minimum thickness required to meet specific property requirements such as wear and corrosion resistance is obtained. This monophase layer is strongly adherent with good lubricity characteristics and need not be removed from components such as crankshafts or gears prior to being used in service.

Plasma nitriding, in common with conventional gas nitriding, increases the fatigue strength of steels. Some data on a 3%CrMo steel is shown in Fig. 5.18[24] and the corresponding hardness profiles in Fig. 5.19.[24]

5. Nitrocarburising

Nitrocarburising, which is usually carried out about 570°C, involves the diffusion of mainly nitrogen, but also some carbon, into the surface with the formation of a thin ($\sim 20\ \mu$m) adherent compound layer with good anti-scuffing properties. This layer consists predominantly of ϵ-carbonitride, the formation of which is favoured by the higher processing temperature (compared to nitriding) and the presence of carbon (Fig. 5.20).[25] Beneath the surface layer there is a diffusion zone up to 1 mm thick, containing finely dispersed nitrides. If after treatment the parts are quenched in warm water or oil, most of the nitrogen in the diffusion zone is retained in solid solution thereby increasing the hardness and the fatigue strength of the material. The presence of a small amount of porosity in the compound layers helps to retain oil under marginally lubricated conditions. Nitrocarburising also improves the resistance of ferri-

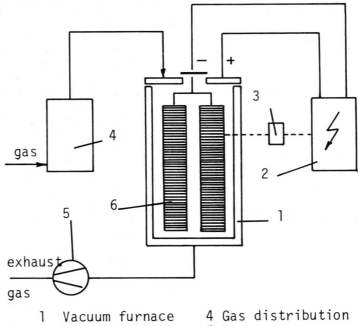

1	Vacuum furnace	4	Gas distribution
2	Electric unit	5	Vacuum pump
3	Regulating device	6	Workpiece

Fig. 5.15 Diagram of ion-nitriding equipment.[22]

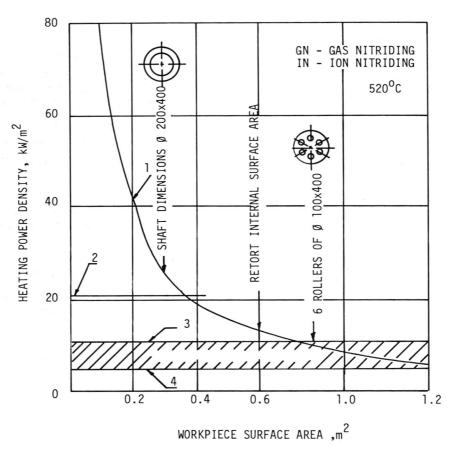

Fig. 5.16 Comparison between heating power consumptions per workpiece surface unit area in ion and in gas nitriding equipment. Temperature is maintained at 520°C.[23]
1 Gas nitriding in a PEH-1 B furnace (retort dimensions ø 300 × 500).
2 Ion nitriding—theoretical upper limit of power consumption for temperature of 520°C.
3 and 4 Ion nitriding—typical power consumption levels.

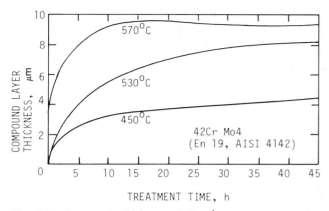

Fig. 5.17 Increase in thickness of the γ^1 compound layer with treatment time for the steel 42Cr.Mo.4 (En19) using ion nitriding at 450, 530 and 570°C.[22]

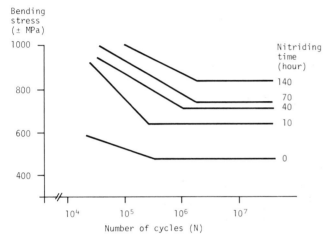

Fig. 5.18 S-N curves of 3%Cr.Mo steel bright plasma nitrided at 480°C.[24]

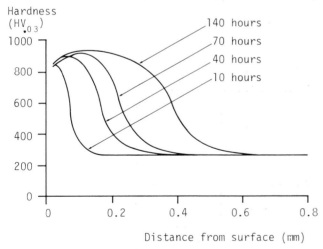

Fig. 5.19 Microhardness profiles of 3%Cr.Mo steel bright plasma nitrided at 480°C.[24]

tic steels and cast irons to corrosion in water and neutral salt solutions.

As most applications of nitrocarburising exploit the properties of the ϵ-carbonitride layer and not the hardness of the underlying steel, the process is used on plain carbon steels and cast irons as well as alloy nitriding steels.

The earlier nitrocarburising treatments involved the use of molten cyanide/cyanate baths. The environmental hazards associated with these have however led to the development of alternative low toxicity baths and a variety of gaseous treatments.

Nitrocarburising is mainly used in the motor industry for applications such as camshafts, crankshafts, rocker arms, rocker shafts, valve guides, shock absorber struts and starter pinions. Other important applications are parts of hydraulic pumps and thin section components in, eg cameras. As the process offers a relatively cheap method of improving the performance of low carbon and low alloy steels its use has increased considerably over the last few years.

The depth of compound layer and its rate of formation can be increased by nitrocarburising in the range 590–720°C, ie between the Fe/N and Fe/C ferrite/austenite transformation temperatures. The process is then known as austenitic nitrocarburising. In the process an additional zone of austenite is formed immediately beneath the compound layer which can be transformed by tempering at 300°C or by a sub-zero treatment. The process differs from carbonitriding in that the core is not austenitic at the processing temperature and alloy steels are therefore not hardened on quenching. (See salt bath Noskuff process p 60).

(i) Nitrocarburising Processes

(a) *Salt bath nitrocarburising*

The earlier nitrocarburising processes, introduced in the late 1940s were based on molten cyanide salt baths. In these the cyanide oxidises to cyanate

$$4NaCN + 2O_2 = 4NaCNO$$

which in turn breaks down according to the reaction

$$4NaCNO = Na_2CO_3 + 2NaCN + CO + 2N_{Fe}$$

Fig. 5.20 Effect of temperature on composition of white layer during nitriding and nitrocarburising.[25]

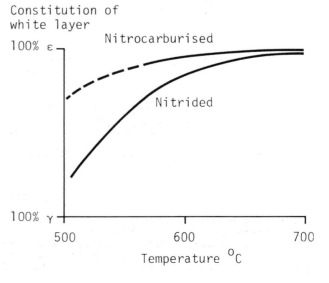

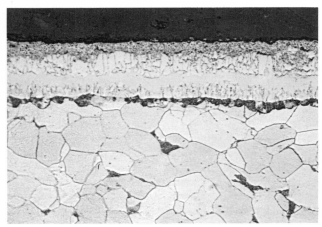

Fig. 5.21 Metallographic section of gaseous nitrocarburised low carbon steel (× 500).

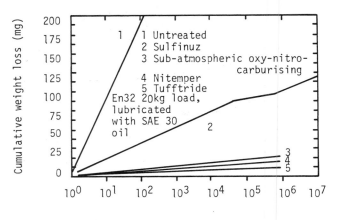

Fig. 5.22 Amsler wear tests on mild steel after various ferritic nitrocarburising treatments.[32]

the steel surface of the components being treated acting as a catalyst.

The two main toxic salt bath treatments of which all others are minor variants are:

(i) The Tufftride or Tenifer process in which dry air is introduced into the bath to stimulate cyanate formation; and

(ii) the Sulfinuz process in which sulphites are added to the bath to aid cyanate formation. Some sulphur is also incorporated in the surface layers.

Variants of the Tufftride process include Mild Nitriding, Activated Nitriding and Aerated Bath Nitriding and of the Sulphinuz process, Sulphidising, Sulfurising and Sulfocyaniding.

In both processes components are preheated to 350–400°C and transferred to the nitrocarburising salt bath at 570°C. Treatment times are usually in the range 1–3 h. To ensure uniformity and reproducibility of the coating, close control over treatment temperature and bath composition must be maintained.

A micrograph, typical of nitrocarburised mild steel, illustrating both the carbonitride compound layer and the diffusion zone in a sample slowly cooled from the treatment temperature, is shown in Fig. 5.21.

Since 1970 concern about the environmental aspects of heat treatment processing with cyanide-based salts has increased. Consequently, a number of salt bath techniques with reduced toxicity levels have been developed. The patented Sursulf[28] process, developed by the Hydromechanique et Frottement Centre in France is marketed for use as an alternative to both the Sulfinuz and Tufftride treatments. In Germany, the Degussa Company has developed an essentially cyanide-free alternative to the Tufftride process called New Tufftride or Tufftride TF1.[29] In each process the toxic cyanide salts have been replaced by thiocyanate which, like cyanides, form cyanates in the bath.

(b) Gas nitrocarburising

A range of gaseous nitrocarburising treatments have been developed based on various gas mixtures, ie

endothermic gas/ammonia;
propane/ammonia/oxygen;
methane/ammonia/oxygen;

impure exothermic gas/ammonia.

The patented Nitemper[30] process involves treatment at 570°C in a sealed quench furnace supplied with a 50% ammonia, 50% endothermic gas mixture. The treatment time is usually 2 to 3 h followed by quenching in oil. The surface compound layer is less porous than that produced by salt bath processes but otherwise exhibits similar properties.

The 'Controlled Nitrocarburising' process[31] is essentially a development of the Nitemper process with closer control of gas composition.

The introduction of a hydrocarbon, in particular methane, to the nitrogen-hydrogen atmosphere of the plasma nitriding process enables carbo-nitride layers to be produced under controlled and reproducible conditions. Although some automobile components are regularly plasma nitrocarburised the process is not yet widely used. A sulphur bearing gas has also been added to the nitrocarburising atmosphere to produce a plasma equivalent of the Sulfinuz process.

Other low pressure processes without the use of a glow dishcarge have also been investigated. For example, treatment for 3 h in a 45%NH$_3$/30%CH$_4$/24%A/1%O$_2$ mixture using a pulsed pressure technique, cycling between 100 and 400 torr, has produced a compound layer essentially similar to that formed by the Nitemper process. Again the process is not widely used.

(ii) Properties of nitrocarburised steels

(a) Tribological properties

The main objective of nitrocarburising is to produce an ε-carbonitride layer which although only 10–20 μm thick markedly improves resistance to adhesive wear (Fig. 5.22).[31] The process is therefore used on many automobile components to aid the 'running-in' process.

(b) Hardness and fatigue strength

In addition to forming the carbonitride layer nitrogen diffuses into the steel to a depth of about a millimetre for normal treatment times. By rapid cooling, it is possible in the case of low carbon steel, to retain

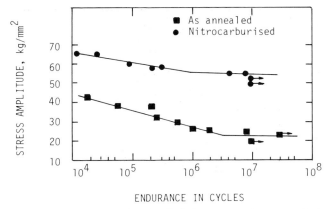

Fig. 5.23 Wohler (rotating bend) fatigue curves for un-notched specimens of untreated and gaseous nitrocarburised mild steel.[32]

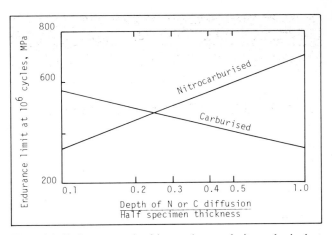

Fig. 5.24 Fatigue strength of low carbon steel nitrocarburised at 570°C or carburised at 900°C.[33]

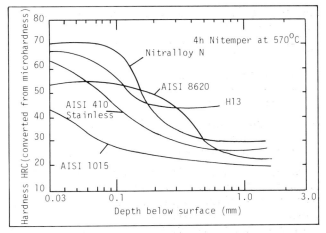

Fig. 5.25 Microhardness profiles of the diffusion zone for a series of steels after the Nitemper treatment.[32]

most of the nitrogen in supersaturated solid solution with a consequent increase in hardness, yield and fatigue strength. Some fatigue results showing an increase of over 100% on nitrocarburising are shown in Fig. 5.23.[33] Unlike the improvements in fatigue strength obtained on carburising and carbonitriding those obtained with nitrocarburising increase with increased depth of nitrogen diffusion zone (Fig. 5.24).

With alloy steels significant hardening is obtained due to the formation of alloy nitride precipitates similar to these produced on nitriding. Hardness profiles for a range of steels after nitrocarburising are shown in Fig. 5.25.[32]

(c) Corrosion resistance

Compared to mild steel, the carbonitride layer is relatively inert to atmospheric corrosion and corrosion in neutral salt solutions. The corrosion resistance of stainless steels is however reduced as a result of nitrocarburising.

A series of salt spray tests have been carried out on mild steel samples to compare nitrocarburising with other coatings. The various samples were,

    untreated and oiled;
    gas nitrocarburised and oiled (17 μm thick);
    phosphated and oiled;
    hard chromium plated (37 μm thick);
    cadmium plated and chromated (12 μm thick).

The tests were carried out for 112 h in a 5% neutral sodium chloride spray. The cadmium plated sample showed no corrosion, the nitrocarburised sample showed only slight patchy corrosion and was significantly less corroded than the chromium and zinc plated samples as well as the phosphated material which were all considerably better than the untreated sample.

(iii) Process developments

In 1978 the Degussa Company of West Germany published a patent demonstrating that a post nitrocarburising salt bath treatment (AB1) could be employed to eliminate the residual toxic effects of a reduced toxic salt bath nitrocarburising treatment (TF1)[35] and in 1980[36] they showed that the treatment also confer-

red a visually acceptable blue/black finish together with significantly enhanced corrosion resistance under saline conditions. Typically the treatment involves preheating of the components in air to about 350°C, followed by salt bath nitrocarburising (TF1) at 580–610°C for between 60–90 minutes. The components are then transferred directly to a second bath based on alkali hydroxide and nitrate (AB1) salt at 330–400°C for between 10–20 minutes which oxidises the compound layer to a blue/black finish. Extensive immersion corrosion tests on a plain carbon 0.45%C steel (DIN 50905) for a variety of treatment sequences are summarised in Table 5.4[37] where it can be seen that a six-step treatment (sequence No. 4) produced the best corrosion resistance. An equivalent gaseous treatment has been developed and marketed by Lucas Industries plc which involves gaseous nitrocarburising in an endothermic gas/ammonia mixture followed by lapping prior to a gaseous oxidising treatment. Both the salt and gaseous processes are currently finding applications in the treatment of symmetrical components such as rams and pistons for hydraulic parts, and are used as direct replacements for hard chrome plating.

A further development by Lucas Industries plc has been aimed at improving the fatigue, scuffing and corrosion resistance of a range of low carbon steel

Table 5.4 Results of extended immersion tests (to DIN 50905 Part 4) on a nitrocarburised 0.45%C steel[37]

Treatment	Weight loss g/m²/24h
1. 90 min TFl at 580°C/water quench	12.3
2. 90 min TFl at 580°C/10 min ABl at350°C/water quench	0.5
3. 90 min TFl at 580°C/20 min ABl at 400°C/ water quench/lap	8.6
4. 90 min TFl at 580°C/10 min ABl at 350°C/ water quench/lap 20 min ABl at 400°C/water quench	0.3
5. 90 min competitive salt at 565°C/water quench	36.3
6. Untreated	39.8

TFl — reduced toxicity salt bath nitrocarburising process
ABl — alkali hydroxide and nitrate salt bath

electrical components for use in automobile manufacture. The recently introduced ultra-low carbon, vacuum degassed and niobium/titanium stabilised deep-drawing steels have enabled many complex thin-sectioned components to be manufactured by single stage press operations. These steels have very low yield strengths of about 155 MPa with elongation values of approximately 45%. However by nitrocarburising followed by quenching from near the treatment temperature the yield strength is increased considerably (Fig. 5.26). Another essential feature of this new patented Nitrotec treatment[38] is that distortion of thin section material can be kept to a minimum by controlled quenching into an oil/water emulsion at a temperature of 70–80°C and that the quench time involved is sufficient (Fig. 5.27)[33] to produce an aesthetically pleasing black oxide film of Fe_3O_4 which needs to be less than 1.0 μm in thickness if exfoliation is to be avoided. Advantage is taken of the microporous nature of the compound layer to introduce an organic sealant to increase the corrosion resistance from 17 to 250 h in salt spray testing. An example of a component treated by this Nitrotec process and tested relative to a zinc plated component is illustrated in Fig. 5.28.[33]

6. Comparison of thermochemical processes involving carbon and nitrogen

(i) Abrasive wear resistance

The hardness of martensitic steels increases with carbon content up to a maximum of about 850 HV at a carbon content in the range 0.7–0.8, the value varying with steel composition. Carburising provides a means of exploiting the wear resistance of high carbon martensites whilst retaining the benefits of the lower carbon tough core. In high load abrasion it is important to ensure that the core is strong enough to support adequately the hardened case. Carbonitriding pro-

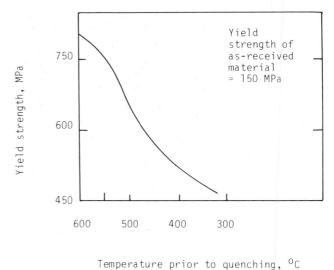

Fig. 5.26 The effect of quenching temperature on the yield strength of nitrocarburised deep drawing steels.[33]

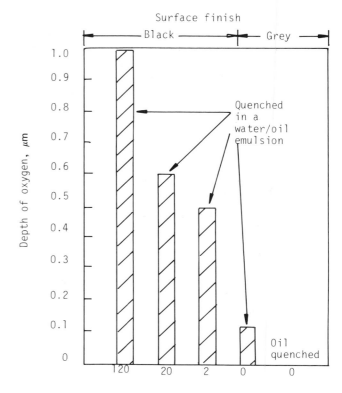

Fig. 5.27 Influence of depth of oxide on surface colouration and corresponding oxidation arrest times for various quench media.[33]

vides cases of similar hardness but, as it is usually performed on plain carbon steels in which the core is relatively weak, it is not suitable for high load applications.

Nitriding provides layers with hardnesses up to 1100 HV on high strength cores and is widely used, eg in machinery for processing filled plastics, for wear resistance.

Nitrocarburised layers have a hardness of approximately 600 HV which provides some resistance to mild abrasive wear. However, the layers (<20 μm

70

Fig. 5.28 Electric fan motor treated by Nitrotec process and neutral salt spray tested for 250 hrs. relative to zinc plated component (left)[33]

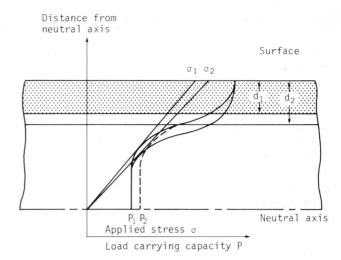

The case depth d_1 on a steel of core strength P_1 can support loads up to a maximum applied stress σ_1. At a higher load, either a steel of higher core strength P_2 or a thicker case (d_2) must be used. (For simplicity the material is considered to be loaded elastically in pure bending and the effects of residual stress are neglected).

Fig. 5.30 Relationship between applied stress, case depth and core strength.[40]

thick) are quickly worn away in conditions of more severe wear.

(ii) Rolling contact fatigue resistance

Rolling contact fatigue resistance is strongly dependent on surface hardness and, as shown in Fig. 5.29,[29] the endurance limit of through-hardened and carburised steels shows an approximately linear relationship to the surface hardness. However, for carburised steels this improvement in fatigue resistance is only obtained when the case depth, hardness profile and core strength are sufficient to prevent premature failure occurring beneath the case (Fig. 5.30).[30] One of

Fig. 5.29 Relationship between surface hardness and maximum compressive stress to produce failure in 10^7 cycles in rolling contact fatigue tests.[39]

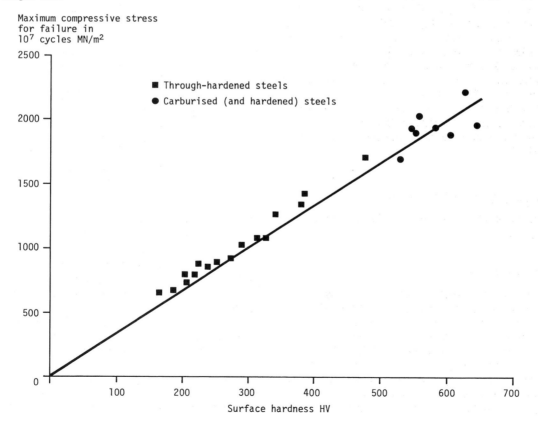

the main advantages of carburising is the ability to produce case depths over a wide thickness range so that the load-carrying capacity can be selected to meet different service conditions. Thermal hardening can also be used to produce thick cases but these are less hard than can be obtained by carburising and therefore the permissible rolling contact fatigue stresses are lower. Nitriding can be used to produce cases harder than those obtained by carburising but in general thinner and for this reason allowable stresses are slightly lower.

By introducing compressive stresses in the surface of components thermal and thermochemical processes increase the fatigue strength of steels. Some data are given on residual stresses and fatigue strengths for various treatments in Tables 5.5 and 5.6 respectively.[41] In components with high stress concentration factors nitriding can be used to increase fatigue strength whereas with carburising there is the danger of quench cracking. Some further data illustrating the improvement in bending fatigue strength of steels on carburising are given in Fig. 5.31.[39]

Table 5.5 Residual stresses measured in surface heat treated steels[41]

Steel	Heat treatment	Residual stress (longitudinal)
832M13 (type)	Carburized at 970°C to 1 mm case with 0.8% surface carbon	(N/mm²)
	Direct quenched	280
	Direct quenched, −80°C sub-zero treatment	340
	Direct quenched, −90°C sub-zero treatment, tempered	200
805A20	Carburised and quenched	240–340
805A20	Carburised to 1.1–1.5 mm case at	190–230
805A17	920°C, direct oil quench, no temper	400
805A17	Carburised to 1.1–1.5 mm case at 920°C, direct oil quench, tempered 150°C	150–200
897M39	Nitrided to case depth of about	400–600
905M39	0.5 mm	800–1000
Cold	Induction hardened, untempered	1000
rolled	Induction hardened, tempered 200°C	650
steel	Induction hardened, tempered 300°C	350
	Induction hardened, tempered 400°C	170

(iii) Adhesive wear and scuffing

Any process which hardens the surface will, by reducing the area of contact between surfaces at a given load, reduce adhesive wear. However the nature of the surface is also important and eutectoid steels in the pearlitic condition, although relatively soft, have good resistance to adhesive wear. The nitride layers developed on nitriding and nitrocarburising prevent metallic surfaces from welding and are therefore beneficial.

Fig. 5.32[32] shows the improvement obtained by carburising and carbonitriding a 0.2% carbon steel in an unlubricated pad on disc test. The wear depth in this test was greater than the layer thickness obtained by nitrocarburising and low temperature carbonitriding (austenitic nitrocarburising) so that the results obtained for these two processes do not reflect the benefits obtained in less severe applications.

Fig. 5.22 illustrates the reduced wear of the various nitrocarburised surfaces obtained in the Amsler test using SAE 30 oil. In this test although the contact stresses are not large, the sliding velocity is high compared with the rolling velocity and boundary lubrication conditions prevail.

(iv) Costs of treatments

Typical sub-contract heat treatment costs prevailing in the UK in 1984 are given in Table 5.7. As can be seen nitriding and nitrocarburising costs are approximately twice the cost of carburising.

(v) Summary

The most appropriate treatment in any particular situation depends on such factors as the type and severity of wear encountered, the configuration and thickness of the parts to be treated, the allowable distortion and the cost. Some of the more important characteristics produced by the four principal thermochemical treatments are summarised in Table 5.8.

Where thick cases are required on heavily loaded components, carburising is used. It is capable of yielding case thicknesses up to several millimetres on carbon and alloy steels with a wide range of core strengths to give a high resistance to abrasion and contact fatigue. The size and shape of components which may be carburised can be limited by the distortion which occurs due to creep at the carburising

Table 5.6 Rotating bending fatigue strength of surface heat treated steels*[41]

	Typical level of fatigue strength (N/mm²)		Change in fatigue strength due to surface heat treatment %	
	Plain	Notched	Plain	Notched
Carburised and quenched	770	500	30–60	100–200
Nitrided alloy steel	770	600	14–30	100–250
Induction hardened	600		20–50	
Ferritic nitrocarburised mild steel	500	330	50–120	80

*These are realistic and typical values that can be achieved by good practice. the notched strengths are achievable with stress concentration factors normally encountered, ie up to about 3

Table 5.7 Typical sub-contract heat treatment costs

Heat Treatment	Average Contract Heat Treatment Costs (£)
(1) Oil harden and temper	0.20/kg
(2) Gas carbonitride (shallow case, (0.25 mm) single quench)	0.22/kg
(3) Gas carburise (0.75 mm case)	0.30/kg
(4) Vacuum hardening—tool steels	
(i) high speed steel + 2 tempers	1.50/kg
(ii) hot work die steel + 2 tempers	1.50/kg
(iii) cold work die steel + 1 temper	1.25/kg
(iv) En 30B	1.00/kg
(5) Gas nitriding (0.5 mm case)	0.50/kg
(6) Gas or salt nitrocarburising	0.60/kg
(7) Solution treat—stainless steel	1.00/kg
(8) 25 cm Induction run (0.75 mm case depth)	0.20/per run

temperature and to the volume changes accompanying the austenite-martensite phase change.

Where distortion is a problem and design stresses are high, nitriding may be used. Some surface finishing treatment may, however, be required to remove the brittle white layer when gas, but not when ion nitriding, is used. The process requires long processing times and relatively expensive steels, but gives a case of high hardness and improved fatigue strength even in components containing stress raising features.

If parts are relatively small, require only a shallow case and if design stresses are low, then carbonitriding is used. It produces a case of higher hardenability than carburising and can be used on plain carbon and very low alloy steels. It has the added advantage of providing better resistance to tempering than carburising and, owing to the lower processing temperature, results in less distortion. The process is generally used

Fig. 5.31 Relationship between tensile strength and bending fatigue strength.[39]

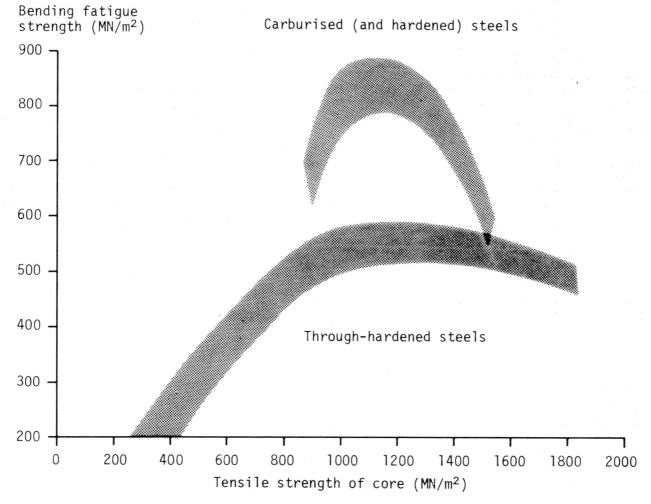

Table 5.8 Factors affecting the application of principal thermochemical treatments

	Carburising	Carbonitriding	Nitrocarburising	Nitriding
Fatigue resistance	✓	✓	✓	✓
Load carrying capacity	✓			✓
Abrasion resistance	✓	✓		✓
Adhesive wear resistance		✓	✓	✓
Suitable for high temperature applications				✓
Gives low distortion			✓	✓
Suitable for cheap steels	✓	✓	✓	
Suitable for large components	✓			✓
Corrosion resistance			✓	✓

to increase the performance of plain carbon steel pressings and stampings.

When adhesive wear is a potential problem, as in the running-in of mating components, the various nitrocarburising treatments can be used. Treated surfaces have a low coefficient of friction and maintain their properties for much longer times than phospha-ted surfaces. Nitro-carburising treatments also improve the fatigue strength and the corrosion resistance of plain carbon and low alloy steels. The treatments are in general unsuitable for high strength steels as the processing temperature (570°C) can soften many steels which have been quenched and tempered.

Fig. 5.32 Wear characteristics of a 0.2% carbon steel both untreated and after different thermochemical treatments.[42]

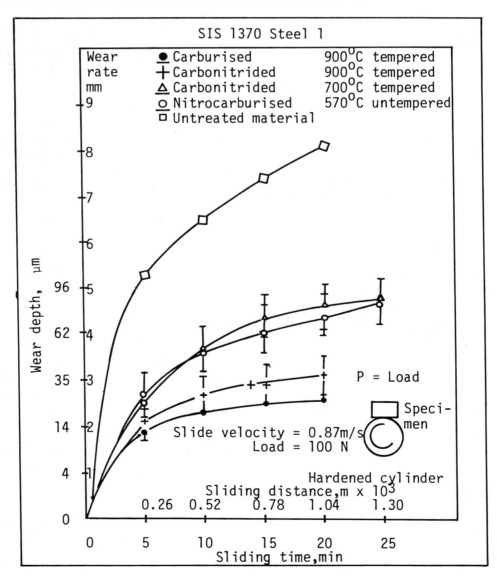

7. Boriding

Boriding, also known as boronising, is a thermochemical diffusion technique whereby boron is diffused into a metallic substrate, usually steel, to form a metal boride layer. Such boride layers, usually single or double phased, have many attractive properties including high melting point and hardness (~1500–3000 HV), good wear and corrosion resistance, and resistance to attack by molten metals.

Boriding treatments are commonly carried out at temperatures in the range 700–1000°C for several hours, producing layers 50–150 μm thick. Layers of greater thickness can be obtained but because of their friability their application is restricted to light contact load situations. Boriding can cause thermal distortion to components of thin section and when precision components are borided, account of dimensional increases (typically 20 to 25% of the boride layer thickness) should be allowed for. Hence, an axle with a 100 μm boride layer will have its diameter increased by 40 to 50 μm. Partial removal of a boride layer, for tolerance requirements, is only possible by diamond lapping which is costly and time-consuming.

There is no appreciable strengthening in the diffusion zone beneath the boride layer. Therefore, components should have sufficient core strength to support the thin hard layers and consequently, boriding is most suited to medium carbon and low or medium alloy steels, or precarburised steel. For low hardenability steels it is necessary to develop core strength by hardening and tempering in subsequent operations, utilising furnaces with protective atmospheres or salt baths, since exposure of the layer to atmospheric oxygen must be avoided.

In those situations where abrasive wear is a particularly difficult problem the high hardness achievable through the boriding thermochemical treatment can be of value (Table 5.9).[43] The process, especially the pack boriding treatment is extensively used in the USSR and in West Germany. However, applications in the UK are restricted, the main reasons for this being a lack of design experience with the treatment, and a lack of boronising outlets.

The high hardness of boride layers has been used to provide wear protection to the steel moulds and extruder screws used in the manufacture of plastics containing abrasive fillers such as mica, TiO_2 and Al_2O_3. The resistance of the Fe_2B boride layer to

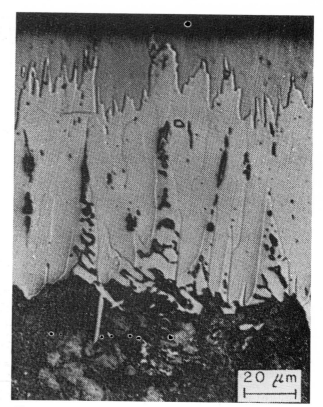

Fig. 5.33 Micrograph of boride layer on mild steel.

corrosion by HCl fumes, released during the decomposition of PVC is an added advantage in such applications. Corrosion resistance of alloy steel pipes used for the transport of vinyl chloride monomer (containing traces of hydrogen chloride vapour) have been enhanced by boriding, thus extending service life by a factor of four.

(i) Morphology of boride layers

The boride layer formed on pure iron consists of:

(a) an exterior layer of orthorhombic FeB; and
(b) an adjacent interior layer of body centred tetragonal Fe_2B.

Both FeB and Fe_2B exhibit a preferred orientation and have a characteristic columnar structure normal to the surface (Fig. 5.33).

The coefficient of expansion of Fe_2B is less than that of iron and hence, upon cooling, this phase remains in compression whereas the coefficient of expansion of FeB is greater than Fe_2B or iron and this phase therefore remains in tension. This disparity in residual stress can result in the formation of cracks in the region of the FeB/Fe_2B interface, especially if the surface is subjected to thermal and/or mechanical shock. Hence, FeB formation should be avoided wherever possible.

Most alloying elements in steel inhibit boride layer growth and the proportion of FeB in the layer increases with alloy content (Fig. 5.34).[44] Such materials as stainless steel are therefore unsuited to boriding, the layers formed being almost, if not

Table 5.9 Comparison of the surface hardness of boronised steels with other representative materials[43]

Material	Hardness (HV)
Tool steel	750
Carburised steel	900
Nitrided steel	1200
Tungsten carbide + 20% cobalt	1000
Tungsten carbide + 13% cobalt	1300
Tungsten carbide + 6% cobalt	1700
Boronised mild steel	1600
Boronised AISI H 13	1800
Boronised AISI A2	1900

entirely, FeB and consequently poorly adherent. Steels containing high proportions of silicon or aluminium should also be avoided since both elements are displaced ahead of the developing boride layer forming a ferrite stabilised region which remains soft after subsequent hardening operations.

(ii) Boriding processes

(a) Pack boriding

This technique is most widely favoured in Western Europe probably because of its relative safety and simplicity. The primary source of boron is boron carbide (B_4C), the boriding capacity of which is enhanced by adding 'activator' compounds such as BaF_2, NH_4Cl or KBF_4. Typical commercial boriding powders, (eg 'Ekabor' and 'Durborid') contain ~5 wt% B_4C, 5 wt% KBF_4 and 90 wt% SiC, the latter material (sometimes substituted by Al_2O_3) acting as an inert diluent. As the B_4C content of the supplied granulated powder is raised the boriding capacity increases and the amount of FeB constituting the layer is increased (Fig. 5.34).[44] Pack boriding can be successfully conducted without solid phase activators by using various reactive gases.

(b) Paste boriding

Paste boriding is only practised when pack boriding is not feasible, eg when certain inaccessible features need to be borided or when masking off is too expensive, difficult or time-consuming. Commercial pastes are supplied in a viscous state and water is added to obtain a consistency which allows application by brushing or spraying. Several coatings of paste are necessary to provide adequate boriding. Components are either heated inductively or in conventional furnaces. In both cases protective atmospheres are necessary. Boriding with pastes is a very labour intensive task which, with the high cost of paste ~£20/kg, often makes the technique prohibitively expensive.

(c) Salt bath boriding

The two techniques used are (a) thermochemical and (b) electrolytic. Both techniques must be used above 850°C to reduce the viscosity of the salts which otherwise impair heat and mass transfer. After treatment adherent salt and excess boron has to be removed which is time-consuming and difficult. Despite high maintenance costs the techniques are widely practised in Eastern Europe, the USSR and Japan.

(d) Gas boriding

Gas boriding is carried out at atmospheric pressure using a 75/25 nitrogen/hydrogen carrier gas with about 0.5 vol% BCl_3. The technique can be used to combine boriding and hardening in one operation.

Plasma boriding is not yet commercially practiced but is the subject of research in Poland, France and the UK. Diborane diluted in hydrogen has been tried and boride layers have been obtained on low alloy steel at 600°C. However, in view of safety considerations, plasma treatments using BCl_3 or other boron halogens, appear to hold most promise.

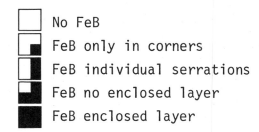

STEEL	B_4C, %			
	2.5	5	7.5	10
C15				
C45				
42Cr Mo4				
61Cr Si V5				
C100				
100 Cr6				
145 Cr6				
x40 Cr13				
18/8				

- □ No FeB
- FeB only in corners
- FeB individual serrations
- FeB no enclosed layer
- ■ FeB enclosed layer

Fig. 5.34 Effect of steel composition and B_4C content of pack on FeB content of borided layers.[44]

(iii) Boriding cemented carbides

Cemented carbides (tungsten carbide-cobalt) have also been successfully borided in commercial boride powders. The treatment produces CoB, Co_2B and Co_3B compounds in the surface region. Boron also replaces some carbon in the carbide, producing $W_2(C,B)_5$ or $W(C,B)_2$. In this application the boriding powder employed contains additional amounts of B_4C to that used for steel treatments. This is essential if the inhibiting effects of a continuous network of tungsten carbide grains on boron diffusion are to be overcome.

This type of treatment has been used on wire drawing dies when lifetimes have been improved by a factor of 8. Drawing speeds have been improved by 50% when drawing high carbon steels and some improvement has been observed when drawing stainless steel, titanium and copper. Despite these promising results borided carbide dies are not in general use.

8. Toyota diffusion (or TD) process

In this process, developed in the Toyota Central Research and Development Laboratories, a smooth

Treatment on Core Pins	1972 Salt bath nitro-carburising	1979 Vanadium carbide coating
Number of casting machines	10	16
Life of pins (shots)	2000–50,000	10,000–150,000
average	12,000	50,000
Number of core pins replaced per machine	1040	290
Frequency of repair work on pins plus die resetting per machine	54	20
Time required for repair work on pins plus die resetting per machine (hr)	56.5	12.5
Time required for polishing pins without die resetting per machine (hr)	115.8	29.0
Time required for polishing per day per machine (mins)	30	7.5

and uniform layer of vanadium, niobium or chromium carbide, a few microns thick, is formed on the surface of a steel by immersion in a fused borax salt bath containing the appropriate ferroalloy. The carbide layer forms by reaction of the metal dissolved in the salt bath with carbon in the steel.

The salt bath temperature is adjusted to the hardening temperature of the steel being processed. For the die steel—Hl3 it is in the range 1000–1050°C. After immersion in the salt bath the component is oil or air quenched and tempered. The rate of formation of a NbC layer on Hl3 steel is shown in Fig. 5.35.[45] Hardness values of 3500 HV and 2800 HV have been obtained at room temperature on vanadium and niobium carbide layers respectively. These values drop to 1400 HV at 600°C and 800 HV at 800°C.

Because of the high processing temperature steps must be taken to minimise distortion as subsequent machining is not possible. The performance of TD treated and nitrocarburised core pins used in die casting of aluminium carburettors are compared in Table 5.10.[45] The process has found widespread use in Japan mainly in dies for sheet metal forming, cold forging and casting (see Chapter 3, p 40).

9. Thermochemical treatment of non-ferrous materials

Some treatments, which involve the diffusional addition of an alloying element (metallic or non-metallic), during a heat treatment cycle can be applied to copper, aluminium and titanium base alloys. The aim of these treatments is to produce a compound layer at the surface of the workpiece under treatment, which is capable of providing wear and/or corrosion resistance. The formation of oxide layers on aluminium and titanium by anodising is discussed in Chapter 6.

Fig. 5.35 Rate of formation of NbC on H13 steel in TD process.[45]

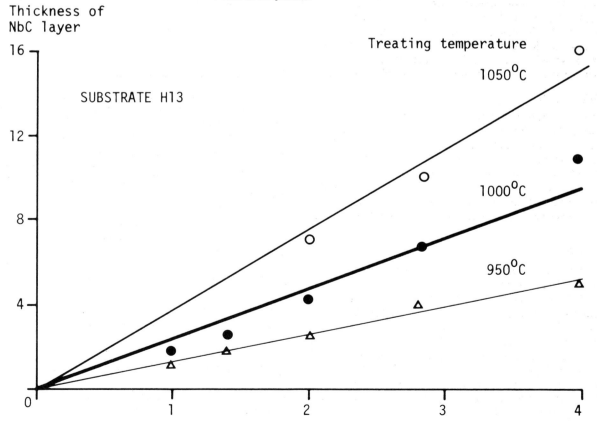

(i) Copper alloys

In the Delsun process[46] a layer consisting essentially of tin is electroplated onto the material under treatment, usually brasses, bronzes or aluminium bronzes. This operation is followed by a diffusion treatment in the range 400°C to 450°C which produces a thin (~15 μm) scuff resistant layer above a thicker hardened layer. Layer hardnesses depend on the substrate but usually lie in the range 450 to 600 HV. The presence of these layers improves both the adhesive and abrasive wear resistance of the surface and enhances the lifetime of components operating in air or water without adequate lubrication. When treated parts are to be used in oxidising environments the inclusion of cadmium in the deposited coating has been found advantageous. The Delsun treatment also improves the corrosion resistance of brasses and bronzes in the standard salt spray test.

Typical applications include valves, gears and pistons, wheels for reduction gears and synchroniser rings for lorry gearboxes.

(ii) Aluminium alloys

In the Zinal process, which is aimed at improving the friction characteristics of aluminium alloys, a coating of an indium alloy containing zinc and copper is electro-deposited onto the surface. A low temperature diffusion treatment (120°C to 150°C) is then carried out. The case consists of two parts; an indium rich outer layer 15–20 μm thick containing 50% indium, 50% copper at the surface above a zone 3–7 μm thick which contains 98% copper. These layers remain sufficiently ductile to deform without cracking and improve scuffing resistance even when lubrication is poor. Test results indicate that Zinal treated aluminium alloys against steel give better results than either hard anodised or hard chromium plated material. The Zinal treatment has no effect on the corrosion resistance. It has been used successfully on diesel and petrol engine pistons, connecting rods, water pump bearings and helicopter rotor parts.

(iii) Titanium alloys

(a) Nitriding

Typical treatments involve heating to a temperature of 800–1000°C in an atmosphere of pure nitrogen or ammonia. Holding at temperatures for up to 16 hours produces a TiN layer approximately 7 μm thick above a diffusion zone which may be up to 100 μm thick.[47] The TiN layer produced has a hardness of about 1500 HV and good frictional characteristics.

More recently nitriding treatments have been successfully performed in a glow discharge plasma, the sputtering action of the glow discharge removing oxide layers and allowing nitriding to occur in a more rapid and uniform manner.

Plasma nitriding markedly improves the friction and wear characteristics of titanium alloys as shown in Fig. 5.36. It also improves corrosion resistance but has an adverse effect on fatigue strength.

Plasma nitriding has been used successfully in the racing car sector of the automotive industry in parts which are subjected to adhesive wear. Typical examples include cam followers and anti-roll bar sockets. A number of gearing components are currently undergoing evaluation for aerospace appli-

Fig. 5.36 Falex test results on untreated and plasma nitrided IMI 318 titanium alloy.[21]

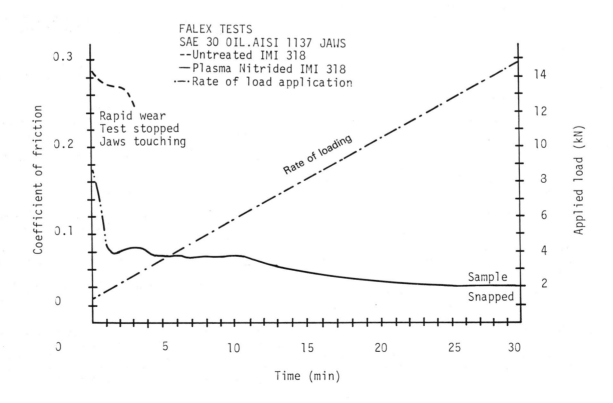

cations. However, the reduction in fatigue strength limits the application of plasma nitriding.

(b) Carburising

Carburised layers have been produced in titanium by heating in propane/argon atmospheres. Treatment times up to 10 hours at temperatures in the range 930 to 1000°C yield surface hardnesses in excess of 1200 HV. Under these conditions a layer about 5 μm thick is produced.[48] Deeper layers can be produced by coating the surface with graphite in a binder and induction heating in a helium atmosphere at 850 to 1100°C. Using this procedure a 250 μm thick layer can be produced in as little as 15 minutes. These coatings have hardnesses > 1500 HV and excellent wear resistance. As in the case of the high temperature nitriding treatments phase transformations may cause distortion of the component.

(c) Oxidation

Enrichment of the surface with oxygen occurs more rapidly than enrichment with nitrogen. The process is usually carried out in salt baths containing mixtures of lithium and potassium carbonates at temperatures between 650 and 800°C for 2 to 4 hours. The wear resistance is improved as a consequence of the hard alpha-case, but as with nitriding treatment the fatigue life is reduced.

(d) Salt bath treatments

Titanium alloys may be treated in cyanide salt baths, to produce a diffusion layer rich in carbon, nitrogen and oxygen. Care must be taken with the oxygen content of the bath if excessive titanium loss is to be avoided. Treatment at 800°C for two hours followed by water quenching or air cooling produces an outer layer of titanium oxide and carbon which is black in colour. This can be removed by glass bead blasting without impairing the wear properties. On a Ti-6Al-4V alloy the treatment reduces the tensile strength by 10% and the yield point by 20%; however, these can be restored by ageing at 530 to 570°C for several hours. The resultant surfaces have frictional coefficients of about 0.2 with no reported loss in fatigue properties.[47]

(e) Boriding

Although not carried out commercially, titanium may be successfully borided in BCl_3/H_2 gas mixtures or using high purity amorphous boron in a vacuum environment. Several hours treatment at 1000°C are required to produce layers 50 μm thick, having surface hardnesses of 2500 HV. The layers are usually duplex in character consisting of TiB and TiB_2.

(f) Chromo-siliciding

In this treatment components are packed into a mixture consisting of 50% chromium/50% silicon in a vacuum furnace. Heating for eight hours at 850°C produces a tough layer of $TiCr_2$ which contains hard intermetallic compounds such as Ti_3Si_5, $TiSi_2$ and TiSi. A surface hardness of approximately 1000 HV is obtained and wear resistance is improved.

References

1 WILKINS, D C A. *Heat Treatment and Quality Control of Metals*. Scientific Publishing Co. Manchester, 1980.

2 WATERFALL, F D. Case hardening steels in cyanide containing salt baths *Metallurgia*, Vol 40, p 29, May 1949.

3 DAWES, C and TRANTER, D F. *Heat Treatment of Metals*, Vol 4, p 121, 1974.

4 WYSS, U and SHAKESHAFT, M. Optimisation of drip-feed carburising atmospheres, 45.1, *Heat Treatment '84*. The Metals Society, London, May 1984.

5 McKENZIE, R T. Fluidised bed heat treatments. *Symposium on the Principles and Practice of Furnace Technology*, NEC, Birmingham, Sept. 1982.

6 LUITEN, C H, LIMQUE, F and BLESS, F. Carburising in vacuum furnaces. *Proc. Conf. Heat Treatment*, Birmingham, The Metals Society, May 1979.

7 STAINES, A M. *Plasma Carburising*. PhD Thesis. University of Liverpool, 1984.

8 COOLIGNON, P and RIBET, F. Industrial applications of ionic carburising, Sect. 49.1. *Heat Treatment '84*. The Metals Society, London, 1984.

9 BOOTH, M, LEES, M I and STAINES, A M. Development of a plasma carburising process. *Proc. Con. Ion Assisted Surface Treatments, Techniques and Applications*, University of Warwick, Sept. 1982.

10 SLYCKE, J and ERICSSON, T. A study of reactions occurring during the carbonitride process. *J. Heat Treating*, Vol 2, 1, p 3, 1982.

11 *The Cassell Noskuff process*. ICI Mond Division

12 *Metals Handbook*, Vol 4, 9th edn. ASM, 1981.

13 DAVIES, R, and SMITH, C G. A practical study of the carbonitriding process. *Heat Treatment of Metals*, Vol 5, p 3, 1978.

14 PARANJPE, V G, COHEN, M, BEVER, M B and FLOC, C F. The iron nitrogen system *Trans. AIME*, Vol 188, p 261, 1950.

15 BELL, T BIRCH, B J, KOROTCHENKO, V and EVANS, S P P. Controlled nitriding in ammonia-hydrogen mixtures. *Heat Treatment '73*, p 51, The Metals Society, London, 1975.

16 JACK, K H. Nitriding. *Heat Treatment '73*, p 39. The Metals Society, London, 1975.

17 *Literature on nitriding steels*. Firth Brown.

18 BARKER, R and SMITH, P K. Response to gas nitriding of 1% CrMo Steel. *Heat Treatment '73*, p 51 The Metals Society, London, 1975.

19 Ion Nitriding, Registered Trade Mark, Klockner Ionon, GmbH.

20 JONES, C K, MARTIN, S W, STURGES, D J and HUDIS, M. Ion Nitriding. *Heat Treatment '73*, p 71, The Metals Society, London, 1974.

21 BELL, T, ZHANG, Z L, LANAGAN, J and STAINES, A M. Plasma nitriding treatments for enhanced wear and corrosion resistance, p 164, in *Coatings and Surface Treatments for Corrosion and Wear Resistance*, ed Strafford, K N, Dutta, P K, and Coogan, C. Ellis-Horwood Ltd, Chichester, 1984.

22 EDENHOFER, B. Physical and metallographic aspects of ion nitriding. *Heat Treatment of Metals*, Vol 1 (1), p 23, 1974 and Vol 1 (2), p 59, 1974.

23 MARCINIAK, A and KARPINSKI, T. Comparative studies on energy consumption in installations for ion and gas nitriding. *Industrial Heating*, p 42, April 1980.

24 BELL, T and LOH, N L. The fatigue characteristics of plasma nitrided 3% CrMo steel. *Journal of Heat Treating*, Vol 2, p 232, 1982.

25 SACKS, R and CLAYTON, D B. Nitriding and nitrocarburising in gaseous atmospheres. *Heat Treatment of Metals*, Vol 6 (2), p 29, 1979.

26 DEGUSSA. British Patent No 891,568. 1960.

27 SOC. DES TRAIGARNENTS DE SURFACE. British Patent No 640,536. 1947

28 CENTRE STEPHANOIS DE RECHERCHES MECANIQUE, HYDROMECANIQUE ET FROTTEMENT. British Patent No 1,424,136. 1976.

29 ASTLEY, P. Liquid nitriding—development and present applications. *Heat Treatment '73*, p 93, The Metals Society, London, 1975.

30 J LUCAS INDUSTRIES LTD. NITEMPER. British Patent No 1,011,580. 1966.

31 BELL, T. British Patent No 1,461,083. 1977.

32 BELL, T. *Ferritic nitrocarburising, Source Book on Nitriding.* p 266. ASM Engineering Bookshelf, 1977.

33 DAWES, C and TRANTER, D F. Nitrotec surface treatment—its development and application in the design and manufacture of automobile components. *Heat Treatment of Metals*, Vol 9 (4), p 85, 1982.

34 BELL, T. A Survey of the principles and practice of thermochemical processing. *Surface Treatments for Protection*, Institution of Metallurgists, London, April 1978.

35 KUNST, H and SCONDO, C. British Patent No 1,527,642, 1978.

36 KUNST, H and SCONDO, C. British Patent No 2,056,505, 1980.

37 ETCHELLS, I V. A new approach to salt bath nitrocarburising. *Heat Treatment of Metals*, Vol 8, (4), p 85, 1981.

38 NITROTEC—Trade Mark of Lucas Industries plc.

39 CHESTERS, W T. Surface hardening of large gears. *Jnl Iron and Steel Institute*, Vol 208, No 11, p 982, 1970.

40 THELNING, K W. *Steel and its Heat Treatment.* Butterworths, London, 1984.

41 CHILDS, H C. *Surface Hardening of Steel.* Engineering Design Guide No 37, Oxford University Press, 1980.

42 KIESSLING, L. A comparison of wear and fatigue characteristics of carburised, carbonitrided and nitrocarburised low carbon steel. *Heat Treatment of Metals*, Vol 6 (4), p 97, 1979.

43 BIDDULPH, R H. Boronising. *Heat Treatment of Metals*, Vol 1, (3), p 95, 1974.

44 SAMSONOV, G V, and EPIK, A P. *Coatings on High Temperature Materials*, (ed Hausner, H H). p 7, Plenum Press, New York, 1966.

45 ARAI, T, and IWAMA, T. Carbide surface treatment of die cast dies and die components. *SDCE Paper* No G–T81–092, Cleveland 1981.

46 GREGORY, J C, Low temperature metal diffusion treatments for the improvement and scuffing and wear resistance of ferrous and non-ferrous metal parts. *Heat Treatment of Metals*, Vol 5 (2), p 33, 1978.

47 WATERHOUSE, R B and WHARTON, M H. *Industrial Lubrication and Tribology*, p 56, March/April 1974.

48 GRIEST, A J, PARIS, W N and FROST, P D, US Patent 2,892,743, 1959.

49 CHATTERJETT-FISCHER, R, and SCHAABER, O. Boriding of steels and non-ferrous metals. *Heat Treatment '76*, p 27, The Metals Society, London, 1976.

Additional reading

Carburising and Carbonitriding. ASM publication 1977.

PARRISH, G. Influence of microstructure on properties of case carburised components. *ASM publication* 1980.

Chapter 6

Electrochemical treatments

1. Introduction

Electrochemical treatments include the electro-deposition of metals and alloys and the production of conversion coatings by anodic oxidation (ie anodising). Electrodeposition from aqueous solutions is a versatile technology with a wide range of applications which has been practised industrially for about 150 years. The use of fused-salt electrolytes extends the range of possible coatings, but the complexity and cost of such processes hinder their general use. Various metals form anodic films, but only aluminium and its alloys are anodised on a large scale.

Chromium, because of its hardness and low friction coefficient against steel, is the most important wear resistant electrodeposit. Nickel is also widely used, especially as an undercoat to chromium in applications requiring additional protection against corrosion. Copper sometimes serves as a cheap undercoat. Cobalt and its alloys have application in coating metal forming dies. Gold and silver are mainly employed in electrical applications, while lead alloys are commonly electrodeposited as overlays on plain bearings. Engineering components made of aluminium alloys are frequently anodised under conditions which yield a thick dense coating (hard anodising). Some composite coatings, comprising powder particles incorporated in an electrodeposited matrix, exhibit exceptionally good wear properties.

Electrodeposited metals tend to be finer grained, harder and more brittle and sometimes more wear resistant than wrought metals. Many deposits have internal stresses, but these can usually be kept to low levels by appropriate choice of process conditions. Thick chromium deposits, however, because of high stress and negligible ductility, always contain fine cracks making the coating permeable.

It is possible to apply electrochemical treatments selectively without complete immersion of the workpiece. The technique is usually referred to as brush plating. It is especially useful for *in-situ* reclamation of large components.

2. Overview of properties and applications

The properties and applications of various electrodeposited coatings are summarised in Table 6.1. Comparative wear data for coatings applied to a 50 mm dia shaft, rotated under a 8.9 N load at a sliding speed of 0.51 m/s in contact with a wear block of Type 440C stainless steel or Type C2 cemented carbide, are displayed in Figs. 6.1[1] and 6.2[1], respectively. The coatings include; as-deposited and hardened elec-

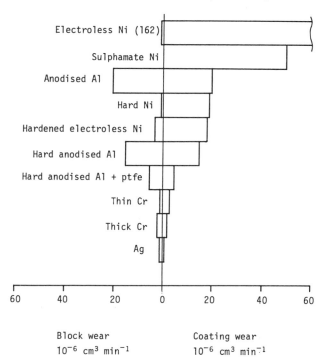

Fig. 6.1 Wear rate of coatings sliding against a hardened Type 440C stainless steel counterface (58–60 HRC).[1]

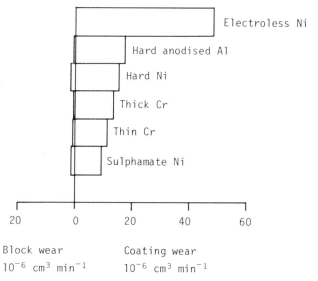

Fig. 6.2 Wear rate of coatings sliding against a Type C2 cemented carbide counterface.[1]

troless nickel; sulphamate nickel (a conventional, machinable deposit); hard nickel from a modified sulphate-chloride bath; conventionally anodised aluminium; hard anodised aluminium, both with and without PTFE impregnation; thin chromium (2–4 μm); thick chromium (100–125 μm); silver. The electrodeposited nickel, thick chromium and silver coatings were ground after plating to a surface roughness of 0.5 μm CLA. The results shown are for low load tests without lubrication. Electroless nickel performs better in relation to electrodeposited chromium in high load Falex tests in the presence of a lubricant, as may be seen in Table 7.3.

Fig. 6.3.[1] gives the results obtained when similar coatings applied to blocks of rectangular cross-section were submitted to a sand/wheel test (ASTM G65). The rubber wheel speed was 2.35 m/s and the abrasive was 50–70 mesh dry sand flowing at 8.8 g/s. The normal force on the specimen had the non-standard value of 5.5 N. The wear rates of the coatings in the abrasive wear test show a roughly inverse relationship with hardness, whereas in the sliding wear tests other factors are significant. For example, the low wear rate of silver is no doubt due to it acting as a solid lubricant with film transfer to the stainless steel counterface. It is also noticeable that the alumina film on anodised aluminium is sufficiently hard to wear the stainless steel but not the tungsten carbide counterface. Electrodeposited chromium shows a low wear rate in both the sliding and abrasive wear tests.

3. Electroplating from aqueous solution

In electroplating from aqueous solutions the article to be plated is made the cathode in a low voltage electrolytic cell. The anodes, usually made of the metal to be deposited, dissolve during the passage of current. Sometimes insoluble anodes are employed (eg lead is normal in chromium plating), in which case oxygen evolution is the principal anode process, and the dissolved metal content must be replenished by suitable additions.

In its simplest form metal deposition may be represented as:

$$M^{n+} + ne^- = M$$

Even when the metal atom forms part of a negative complex ion, rather than existing as a simple (or hydrated) positive ion, it is still reduction by electrons that produces the metallic phase. If the potential at which the metal is deposited is moderately more negative than that at which hydrogen ions can be reduced on that surface, then metal deposition and hydrogen evolution occur simultaneously. Thus nickel is deposited at a cathodic current efficiency of about 95% whilst chromium generally deposits at efficiencies below 20%. Metals with electrode potentials which are more than about 1 V more negative than the standard hydrogen potential, eg titanium and aluminium, cannot be electrodeposited from aqueous solution.

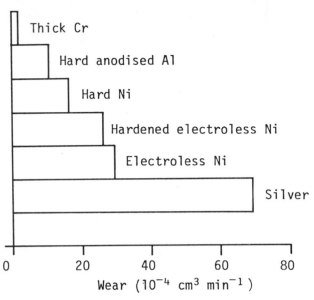

Fig. 6.3 Wear rate of coatings in the dry sand/rubber wheel test (ASTM G65).[1]

(i) Equipment and Process

(a) Process tanks

The material of construction is governed by the nature of the solution and the operating temperature. Glass fibre reinforced plastic tanks are widely used for rinses and dips, but lined steel tanks are most common for the principal operations. Natural and synthetic rubbers and various plastics—pvc, perspex, polypropylene, polyethylene—are the most usual lining materials. Antimonial lead linings, with inner loose linings of wire reinforced glass or plastic, which used to be standard for chromium plating and anodising, are being replaced by suitable plastics.

(b) Heating and cooling

Heating may be by steam or high pressure hot water, electric immersion heaters or directly by gas. Heating coils or heat exchangers of titanium are suitable for both nickel and chromium plating solutions. Water jackets are usual for heating lead-lined chromium plating tanks. Hollow plastic balls are often floated on the surface of plating solutions to conserve heat, as well as to reduce spray and evaporation. Solution cooling is always required for hard anodising and may be necessary in plating tanks if they are operated continuously at high load. Titanium coils or heat exchangers are best for this purpose.

(c) Agitation and filtration

Agitation is used to keep solutions well mixed and allow plating to proceed at high current densities. The preferred method is by low pressure, oil-free compressed air, introduced via perforated pipes placed near the bottom of the tank. Alternatively a reciprocating mechanical motion may be employed if the bath contains a wetting agent or oxidisable constituents. No agitation is necessary in chromium plating solutions because of the large volumes of gas evolved at the electrodes.

Continuous filtration is desirable to avoid roughness

Table 6.1 Properties and applications of some electrodeposited coatings

Coating	Thickness, μm	Hardness, HV	Friction coeff. v steel	Taber wear index	Applications
Chromium	Up to 500	850–1000	0.16	2	Hydraulic rams, exhaust valve cylinder liners, extruder screws, moulds, dies, cutting tools, rotors, reclamation.
Nickel	Up to 500	200–300 (Up to 700 with additives)	0.26 (lubricated)	25	Moulds, hydraulic parts, food machinery, fuel pump and oil well components, airscrews, reclamation, screen printing cylinders, record stampers.
Ni-Co	Up to 125	Up to 500	—	—	Moulds for glass and zinc die castings, electrotypes.
Copper	Up to 500	60–150	—	—	Drawing lubricant, carburising resist, release agent for glass printing rolls, undercoat in reclamation.
Cobalt	Up to 500	180–440	0.45	—	Used in past for record stampers.
Co–W, Co–Mo	Up to 100	350–850	—	—	Hot working dies.
Iron	12–24	150–350			Cylinder coating for ic engines, printing plates.
Gold + alloys	Up to 3	65–385	—	—	Electrical contacts.
Silver + alloys	Up to 10*	60–150	0.16–0.33	—	Bearings, sliprings, rotary switches.
Pb-Sn, Pb-In, Pb-Sn-Cu	25	8–15			Overlays on plain bearings.
Ni-SiC	Up to 250	—	—	—	Cylinder coating for ic engines.
Ni-Al$_2$O$_3$	Up to 250	430	—	12	Cylinder coating for ic engines.
Co-Cr$_3$C$_2$	Up to 250	465†			Compressor blade roots, leading edge of blades, pump parts.
Borided steel	100	1500–2500	—	—	
Borided Mo	—	4000	—	—	
Beryllides	—	500–2500			
Hard anodising of Al	Up to 50‡	350–600§			Injection moulds, aircraft components, rolls, cams and cam tracks, pump turbines, ball race housings, gun barrels, connector housings.
Hard anodising of Ti	10	500	0.01–0.05 (lubricated)		Aerospace components.

*Very thick coatings are feasible for bearings.
‡100 μm or more possible in certain cases.
†600 after heat treatment.
§Values affected by coating structure; abrasion resistance corresponds to value of about HV950.

due to incorporation of suspended particles deriving from the anodes, the atmosphere or the work itself. Plate, cartridge and bag type filters are used, with various filter media. Centrifugal or self-priming pumps may be employed, or diaphragm pumps in the smaller installations.

(d) Ventilation

A high level of general ventilation is necessary in plating shops. It is good practice to install air extraction on all tanks operated at temperatures above 60°C. Extraction via lip exhaust is essential on chromium plating tanks, because of the toxic spray they emit, and desirable on acid copper plating and anodising tanks. The spray generated in chromium plating can be substantially reduced by the use of wetting agents that cause a foam blanket to form on the surface of the bath.

(e) Electrical supply

Transformer/rectifier sets produce the necessary low voltage dc supply from the ac mains. Selenium rectifiers are still in use, but most new ones are of the silicon type. An 8 V unit is satisfactory for the majority of plating processes, 12 V for hard chromium, but 60 to 80 V may be needed for hard anodising. Three-phase rectification is preferred, and is essential for chromium plating.

(f) Support of workpieces

Fasteners and other small articles are plated in barrels, with perforated walls, rotating in the plating bath. The anodes are outside the barrel, cathode contact being made inside via the ends of insulated cables (danglers). Handling is simple, there are no contact marks, and rarely any problems with pitting or roughness. Coating thickness, however, is limited to a few microns and cannot be controlled exactly, being determined by a bell-shaped frequency distribution curve. Components which are large, or easily damaged, or require a perfectly smooth surface, or need to be stopped off or shielded, cannot be barrel plated. Most engineering components, therefore, are plated on racks or jigs. These are electrically conducting frames covered with chemically resistant and insulating plastic. A hook at the top engages the cathode bar and spring contacts support the components. Anodising jigs are often made entirely of titanium, which requires no insulating cover.

(g) Surface preparation

Thorough preparation is essential and usually involves degreasing, soak cleaning, electrolytic cleaning and etching. Masking of areas that are not to be coated is often carried out after the early cleaning stages. Insulating adhesive tapes or lacquer may be used. Alternatively, the component may be immersed in molten wax, which after coating is removed from the areas to be plated, these having first been painted with a paste of powdered chalk and glycerine to prevent sticking.

Degreasing is carried out in hot chlorinated hydrocarbon vapour, sometimes preceded by immersion in the boiling liquid. Soak cleaners may contain sodium hydroxide, metasilicate, carbonate, cyanide, tripolyphosphate or hexametaphosphate, plus wetting and sequestering agents.

Electrolytic cleaning is done in alkaline baths similar to soak cleaners, but with current passing to contribute a scouring action by gas generation. Both cathodic and anodic cleaning are used, the choice depending on susceptibility to attack or hydrogen embrittlement and the risk of depositing contaminants. Etching is necessary to expose a clean, firm structure. A dip in dilute hydrochloric acid may be sufficient, but for the strongest adhesion on a steel substrate an anodic etch in sulphuric acid solution is preferred. Special pre-treatment procedures are required for stainless steel, aluminium and titanium alloys and other materials protected by a tenacious oxide film. These procedures involve the removal of the oxide film mechanically using abrasives, or chemically using an appropriate acid and initial plating in a bath which prevents the reformation of the oxide film. In the case of aluminium an initial coat of zinc can be obtained by immersion in a zinc salt solution.

(h) Hydrogen embrittlement

Certain materials, especially high strength steels, absorb atomic hydrogen during processing in aqueous solutions, and become subject to delayed brittle failure at stresses well below the yield strength. To reduce the risk of damage, heat treatments are performed, both before and after plating. The conditions to be used are currently under review; the recommendations below are taken from BS 4758:1971 Electroplated Coatings of Nickel for Engineering Purposes.

Severely cold worked steel, or high strength steel that has been ground or machined after tempering, is stress relieved before plating, normally by heating at 190–230°C for not less than 1 h. After plating, articles of steel with tensile strength below 1000 N/mm^2 require no heat treatment; otherwise they are heated at 190–230°C for a time that increases with strength as follows: 1000–1400 N/mm^2, 2 h; 1400–1800 N/mm^2, 6 h; greater than 1800 N/mm^2, 18 h.

(ii) Chromium

(a) Plating process

Traditionally, very similar processes have been used to deposit chromium for decorative and engineering applications, the main difference being in the coating thicknesses, which are typically much less than one micron and 10–500 μm respectively. With the introduction of micro-cracked and micro-porous deposits and, more recently, trivalent chromium processes for decorative use, a greater divergence has occurred. Chromium for wear resistance (often called hard chrome) is still generally deposited under conditions such as the following:

bath composition	CrO_3	250 g/l
	H_2SO_4	2.5 g/l
current density		4.0 kA/m^2
voltage		6–7 V
temperature		50–60°C
deposition rate		25 $\mu m/h$

Higher efficiencies and faster deposition rates are possible if fluoride or fluorosilicate catalysts are used in place of some of the sulphuric acid, but the solution is then more corrosive and greater care is necessary to protect workpieces from attack. In so-called self-regulating baths, other additions are made to simplify control of the catalyst concentration.

Porous chromium is produced by chemical or electrolytic etching of thick chromium coatings. It has a pit or channel type of porosity, depending on the deposition and etching conditions. Another form is produced by simply plating on to a severely roughened substrate. Because it retains lubricant well, it is used mainly for engine cylinders and piston rings.

(b) Properties

Electrodeposited chromium contains a fine dispersion of chromium oxides and is very fine grained, X-ray line broadening indicating a grain size below 10^{-2} μm. Because of their brittleness and high tensile stress, chromium coatings crack spontaneously at a thickness of about 0.5 μm. Solution drawn into the cracks and covered by the subsequent deposit is probably the origin of the oxide inclusions. Cracking occurs repeatedly in layers and although the successive crack patterns do not generally coincide there are inevitably, some interlinked channels giving a degree of permeability.

It is possible, by doubling the current density and raising the bath temperature to 85°C, to deposit a coating with lower stress and greater tensile strength (about 480 compared with 100 MN/m^2). It is also softer than conventional chromium. Table 6.2[2] shows how the hardnesses of the two deposits compare at different temperatures and at room temperature after

Table 6.2   Hardness of chromium plate at various temperatures[2]

Temp of determination, °C	20	200	400	600	800	20 (after all tests)
Conventional chromium	1000	810	585	270	100	300
Low-stress chromium	620	520	380	240	100	220

the full sequence of measurements. The temperature was raised at a rate of 30–60°C/min and held at each test temperature for 15 min before making the hardness determination.

As-plated chromium deposits suffer a significant loss in room-temperature hardness if they are heated for one hour to 200–250°C or above. If they have previously been subjected to a nitriding or carbonitriding treatment, however, heating to over 600°C becomes possible without the room-temperature hardness falling below 800 HV[3] (see Fig. 3.6).

Chromium has a low coefficient of friction against many of the commonly used engineering metals, as illustrated in Table 6.3.[2] The coefficient of friction of the chromium/cast iron couple is shown in Fig. 6.4[2] as a function of temperature, with and without lubrication.

The rate of abrasive wear of chromium is inversely related to its hardness, as shown in Fig. 6.5;[2] the tests being made on a chromium plated ball pressed on to a rotating emery disc. The wear rates produced by synthetic yarn moving at high speed on chromium coatings, as deposited and after heat treatment at various temperatures, are shown in Table 6.4.[4] The tests were made with a 15 denier, dull monofilament of nylon, running at 914 m/min under 15 g tension.

In a series of Falex tests in which steel pins plated with 50 μm chromium were revolved between low alloy steel V-blocks lubricated with mineral oil, the wear loss was about one fourteenth that of unplated low alloy pins (SAE 3140)[5]. In lubricated metal-to-metal sliding contact, ordinary and porous chromium behave similarly at low loads, but above 2 MN/m² porous chromium has a much longer life.

Table 6.3   Friction coefficients of various couples[2]

Metal couple	Static coefficient	Dynamic coefficient
Cr plated steel/Cr plated steel	0.14	0.12
Cr plated steel/babbitt	0.15	0.13
Cr plated steel/steel	0.17	0.16
Steel/babbitt	0.25	0.20
Babbitt/babbitt	0.54	0.19
Steel/steel	0.30	0.20

Table 6.4   Yarnline wear resistance of chromium plate[4]

Condition	Hardness, HV	Wear rate, μm/h
As plated	1020	0.9
After 100°C for 4h	1010	0.9
After 200°C for 4h	985	1.6
After 400°C for 4h	811	3.1
Test standard:		
Vasco 7152 tool steel	520	2.25

At elevated temperatures, chromium softens, but a hard oxide forms. Under heavy loads, the oxide wears away as quickly as it grows, and the performance, controlled by the hardness of the metal, deteriorates. Consequently, 400°C is a practical limit for high load applications. At low loads, however, the limit is higher, and the performance improves with increase in temperature. Thus, the wear rate against 12% Cr steel and 18% Cr 8% Ni steel at 500°C is only about 1% and 5%, respectively, of the corresponding rates at room temperature. At 600°C, the wear rate of chromium against Nimonic 80A is similar to the value

Fig. 6.4   Coefficient of friction of chromium plated cast iron as a function of temperature, with and without lubrication.[2]

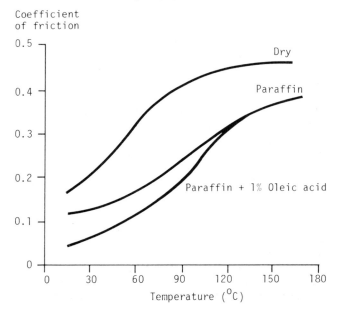

Fig. 6.5   Relation between abrasive wear and hardness of hard chromium plate.[2]

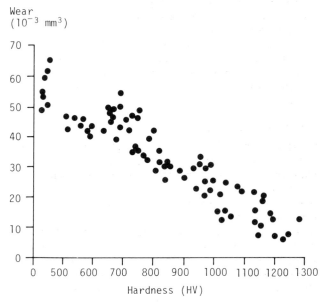

85

at room temperature, and against itself it wears at only 7% of the room temperature rate[6] (Table 10.11).

Electroplating with chromium reduces the fatigue strength of most steels and titanium alloys. With steels, there is a linear relationship between the percentage change in fatigue limit and the fatigue strength of the substrate,[7] as has already been discussed in Chapter 2 (Fig. 2.4). Shot peening before plating substantially reduces the adverse effect of a chromium coating on the fatigue limit of high strength steel. In one case, the 100,000 cycle failure stress fell from 827 to 758 MN/m^2 with peening, and to 414 MN/m^2, without.[8] The loss in fatigue strength suffered by titanium alloys on chromium plating can also be reduced by prior peening. To secure strong adhesion, however, a post-plating heat treatment at 450°C for 1h is recommended. On alloy IMI 680 (4% Al, 4% Mn), this had no further effect on fatigue strength, but on alloy IMI 314 (10.7% Sn, 4.1% Mo, 2.3% Al, 0.2% Si) the fatigue limit fell by 54%, probably because of the relaxation of the compressive stress induced by peening.[9]

(c) Applications

The engineering applications of chromium plate exploit the following properties:

> low coefficient of friction;
> anti-stick capability;
> wear resistance;
> corrosion resistance in certain environments;
> load bearing ability (provided the substrate gives adequate support).

The following thicknesses are recommended for deposits used as-plated or merely polished or honed; up to 12 μm for moulds for plastics and cutting tools; 12–50 μm for hydraulic rams and cylinder liners for internal combustion engines; over 50 μm, for resistance to corrosion and wear where finish and dimensions are not of prime importance. Deposits that are finished to size by grinding should usually have a final thickness of 100–250 μm. If a greater total thickness is required, eg in the reclamation of under-sized components, it is preferable to make up the difference with an undercoat such as nickel.[10]

(d) Current developments

The first commercial chromium plating process based on trivalent salts was introduced in 1975.[11] It has many advantages for decorative plating, yielding coatings that are equal in durability to those from hexavalent chromic acid baths, while avoiding their health hazards and effluent problems. The deposit is darker than normal chromium, especially on textured or satin surfaces, and its maintenance requirements are more stringent. These factors have impeded its general adoption. A further trivalent process, which was fully launched in 1983 after several years' development, is said, however, to make it easier to achieve the desired colour.[12] Using these processes it has not so far proved possible, however, to attain the requisite hardness for engineering applications. In June 1984, the first commercial processes for the electro-deposition of

chromium-iron and chromium-iron-nickel alloys were announced.[13] These are also based on dilute solutions of trivalent chromium. The present recommendation is to use a 1 μm alloy deposit on a 25 μm nickel undercoat for decorative and protective applications.

In some applications, plasma spraying of metals, ceramics or composites competes with chromium plating and has the advantage of lower equipment costs and higher deposition rates.[14] The coatings can be harder and more abrasion resistant and able to resist higher temperatures, but they are more porous, less adherent and have inferior impact resistance. Masking, stripping and surface finishing are also more difficult. In some markets, plasma or flame sprayed molybdenum has taken over from porous chromium as a facing for piston rings (see Fig. 10.11 and Figs. 3.7 and 3.8).

Chromium plating faces competition also from CVD and PVD coatings (see Tables 8.3 and 8.4 and Fig. 9.6 respectively) and from electroless nickel. There is no sign, however, of any major decline in the use made of this long established process.

(iii) Nickel and its alloys

The two following baths are most commonly used in the deposition of nickel for engineering purposes:[15]

Watts	
nickel sulphate (NiSO$_4$.6H$_2$O)	225–375 g/l
nickel chloride (NiCl$_2$.6H$_2$O)	30–60 g/l
boric acid (H$_3$BO$_3$)	30–40 g/l
temperature	45–65°C
pH	1.5–4.5
current density	250–1000 A/m^2
deposition rate	30–120 μm/h

Sulphamate	
nickel sulphamate (Ni(NH$_2$SO$_3$)$_2$)	270–330 g/l
nickel chloride (NiCl$_2$.6H$_2$O)	15 g/l
boric acid (H$_3$BO$_3$)	30–45 g/l
temperature	25–70°C
pH	3.5–4.2
current density	200–1400 A/m^2
deposition rate	25–171 μm/h

Chloride-free sulphate solutions are sometimes also used. Deposits from the above baths are ductile and soft (140–280HV).

Another process using a highly concentrated solution of nickel sulphamate operated at 60–70°C has been shown to yield sound deposits at exceptionally high current densities. As with many nickel plating processes, continuous conditioning, i.e. low current electrolysis of the solution in a separate tank is recommended to remove impurities. With this high concentration sulphamate process, close control of the internal stress and hardness (up to 400HV) of the deposit by choice of operating temperature and current density has been demonstrated.

Bright nickel deposits, produced for decorative purposes using organic addition agents containing sulphur are much harder, up to 600 HV, but generally

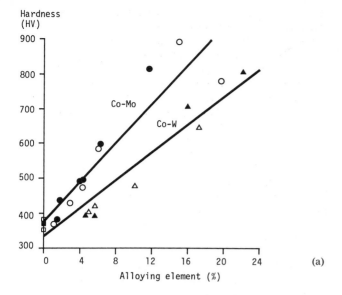

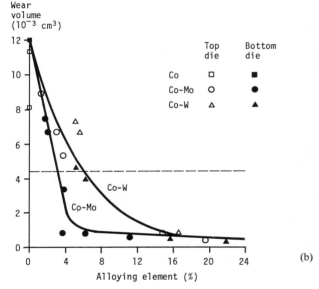

Fig. 6.6 Properties of cobalt alloy plated dies.
(a) Hardness of cobalt alloys as function of W and Mo contents.
(b) Volume loss of alloy after forging 1000 billets. (Dotted line shows mean volume loss of unplated No. 5 type die steel).[23]

unsuitable for engineering use. It is possible to modify the bath composition without adding organic substances (ammonium salts are sometimes used), increasing the hardness to 400 HV or more. A similar hardness can be attained using an organic, but non-sulphur containing addition, benzamide.[16] In contrast to chromium, electrodeposited nickel tends to gall when rubbed against a steel or another nickel surface. It is therefore mainly used when enhanced corrosion resistance is required, often as an undercoat for chromium.

It is not unusual for nickel to be deposited as very thick coatings—up to 500 μm on new parts and perhaps 5 or 6 mm on salvaged components. These coatings can be finished either by grinding or machin-

ing. Electroforming in nickel is widely practised, the products produced on the largest scale being screen printing cylinders and record stampers.

Nickel-cobalt alloys are readily deposited from concentrated sulphamate solutions. With a cobalt content of 33.5% a hardness of 500 HV can be attained which falls to 400 HV after 17 h at 400°C. Because of the absence of incorporated sulphur, the coating does not embrittle on heating. Under practical operating conditions an alloy with a cobalt content of 15% and a hardness of about 380 HV, which falls to 210 HV after 17 h at 400°C is used.[17] The process has been used successfully for electroforming zinc die casting dies.

(iv) Copper

Copper is electrodeposited from a wide variety of solutions. Cyanide solutions have very good covering and throwing power, and can be used to deposit directly onto steel or zinc. Pyrophosphate, sulphate or fluoborate baths are capable of very high deposition rates and are used for depositing thick coatings.

Depending on the deposition conditions the grain structure of electro-deposited copper can vary from coarse to fine and the hardness from 60–150 HV. Copper, sometimes with a top coat of nickel and/or chromium, is used for reclaiming damaged or worn surfaces if these are only subject to low loads. It is used on cylinders and plates for printing by the intaglio process. A layer of chromium is sometimes put on top to provide extra wear resistance or protection against the variety of inks used in colour printing. Copper is sometimes used as a lubricant in the drawing of steel wire and as a release agent on the surface of moulds. It is the standard coating applied to protect selected areas of components from the effects of a case-hardening treatment.

(v) Cobalt and its alloys

Cobalt and cobalt-nickel alloys can be deposited from the same types of solutions as nickel. Wear resistant alloys include cobalt-molybdenum and cobalt-tungsten, electrodeposited from a sulphate bath containing sodium heptonate plus sodium molybdate or tungstate,[18] and cobalt-rhenium, deposited from a sulphate bath with additions of ammonium perrhenate and perrhenic acid.[19]

Cobalt and cobalt alloy, with up to 22% nickel, deposits have a columnar microstructure. Alloys with more nickel and those containing tungsten have a banded structure, recrystallisation occurring at 600°C for the nickel and at 730°C for the tungsten alloy. Cobalt-rhenium alloys have a mixed laminar/columnar structure and recrystallise at 980°C. The lattice of Co, Co-W and Co-Re is hexagonal close packed, but that of the alloy with more than 30% nickel is face centred cubic.

Pure cobalt from a sulphamate bath has a hardness between 250 and 330 HV. It softens at temperatures above 200°C, falling below 100 HV at 600°C; on subsequent cooling, however, the hardness reverts to a value in the region of 350 HV. Alloys with 60%

nickel have a hardness of 500 HV initially; these also soften on heating but there is no substantial recovery on cooling.[20] Co-W and Co-Mo alloys have hardnesses up to 800 or 900 HV, depending on composition, and do not soften greatly at temperatures up to 600°C. Co-Re has a hardness in the range 400–750 HV, raised by heat treatment at 330°C to 770–1200 HV, and maintained at temperatures up to 815°C.

No wear tests on electrodeposited cobalt have been reported, but wear of a solid cobalt pin against a rotating En3A steel disc was only one thirtieth that of an En3A pin in the same test.[21] The coefficient of sliding friction between cobalt and itself in air is 0.37, increasing to 0.65 above 417°C, where the structure changes from hexagonal close packed to face centred cubic.[22] Cobalt is appropriate for use in space applications because, unlike other metals, its friction coefficient and wear rate do not increase under high vacuum. However, pure cobalt has no current industrial applications as a coating, except as a matrix for a composite electrodeposit.

Co-W and Co-Mo coatings, with thicknesses in the range of 25–100 μm, were applied to shaped dies, which were used to forge a thousand En3B steel billets. The degree of deformation was increased and the die temperature raised from 130 to 250°C to make the test a severe one. Fig. 6.6 shows the wear volume and hardness of the coatings plotted against composition. The fact that neither of these alloys scaled during the test helped in reducing wear. Further development of these alloys as die coatings has been in the context of brush plating.

(vi) Iron and its alloys

Iron, and iron-nickel alloys, can be deposited from solutions based on ferrous chloride, sulphate, sulphonate, sulphamate or fluoborate. Oxidation of ferrous to ferric ion can cause problems, which are more easily overcome if the process is operated continuously rather than intermittently. Raising the current density and lowering the operating temperature tends to increase deposit hardness. Additions of relatively simple organic compounds produce further increases in hardness.

The hardness of electrodeposited iron varies over a range from 120 to 500 HV. Harder deposits with little or no ductility are produced by operating at temperatures below 65°C and by adding, as grain refiners, glycol, sugar, formic, acetic, citric or ascorbic acid. The coatings contain included iron oxide or hydroxide. A deposit from a sulphate bath with citric acid additions, deposited at 25–35°C, can have a hardness around 1000 HV and contain up to 0.7% carbon. Deposits of hardness 350–650 HV produced using oxalic acid additions are claimed to resist wear better than chromium. Iron-tungsten alloys can be deposited with hardnesses approaching 1000 HV and heat treatment at 600°C raises this to 1300 HV.[20] Iron-4% nickel coatings applied to bearings can be carburised to a hardness of 700–750 HV and perform better than En31 or En39 bearings in a four-ball test for rolling contact fatigue life.[24]

Iron deposits have been used over a long period

Table 6.6 Properties of pure gold and gold—0.2% Co alloy[28]

Property	Wrought Au	Electrodeposited Au-Co
Density, g/cm³	19.3	17.5
Composition, %	99.99	Au 99, Co 0.2, C 0.2, K 0.2, O, N, H
Electrical resistivity, 10^{-6} Ω-cm	2.3	5–10
Tensile strength, MN/m²	131	207
Elongation, %	45	<0.4
Hardness, HV	25	180

Table 6.5 Hardness and wear properties of gold deposits[27]

Types of deposit	Hardness, HV	Width of wear track, μm
Soft unalloyed gold	65	—
Hard unalloyed gold	200	914
Au-0.2%Co	175	140
Au-18%Cu-7%Cd	385	53

for the reclamation of large components, eg marine diesel cylinder liners, before applying a top coat of chromium. A widely publicised application was on the aluminium pistons used in the aluminium-16% silicon alloy engine of the Chevrolet Vega; a thin final coat of tin was applied to help running in.[25]

Many applications of iron deposition have proved short-lived, the cheapness of the metal failing to compensate for problems with process control, stress, brittleness and rusting. An exception was in the production of intaglio plates for printing US banknotes; these were nickel faced with a heavy backing of electrodeposited iron. Much work has been done in recent years to develop electroforming with iron, mainly for use in producing moulds and dies.[26]

(vii) Gold and its alloys

Unalloyed gold is deposited mainly from near neutral solutions containing potassium and gold cyanides and buffering agents such as phosphates or citrates. The coatings produced vary widely in hardness, depending on the plating conditions, but all tend to fail rapidly by adhesive wear in unlubricated sliding contact applications.

Gold alloys are much more resistant to wear in the absence of a lubricant than pure gold. They are obtained from the same type of solution, with additions of soluble salts or complexes of metals such as cobalt, nickel, silver or copper, the pH being usually adjusted to a value in the range 3.0–5.0. The alloying constituents both harden the deposit and promote the formation of superficial oxide films which impede contact welding.

The hardness and wear behaviour of several gold coatings are reported in Table 6.5. The wear tests were carried out dry, using a 6.35 mm dia. pure gold rider in contact with a plated disc. In Table 6.6, the properties of an electrodeposited gold-0.2% cobalt alloy are compared with those of wrought pure gold.[28]

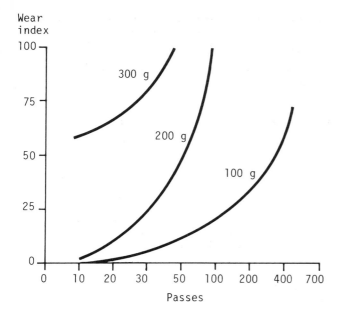

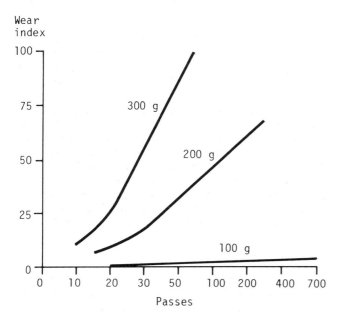

Fig. 6.7 Adhesive wear of Au-Co coating on copper.[28]
(a) Without and
(b) With 2.5 μm Ni undercoat at various loads.

The higher hardness and lower ductility of the gold-0.2% cobalt alloy are responsible for its superior adhesive wear characteristics and much lower coefficient of friction in unlubricated sliding (0.4–0.8 compared with 1.5–2.2).

Gold and gold alloy deposits are soft and thin and their wear resistance is increased by the use of a hard undercoat. The results of tests on 3.3 μm Au-Co coatings on copper, with and without a 2.5 μm nickel undercoat of hardness close to 500 HV are shown in Fig. 6.7.

Because of chemical inertness, high conductivity and low contact resistance, electrodeposited gold and its alloys are used in many applications in electronics,

alloyed deposits of a few microns thickness being chosen for contact situations where wear resistance is necessary. Because of the cost of the metal, it is usual to limit the area coated as far as possible by the use of selective plating techniques. Much effort has gone into the search for cheaper alternatives, the leading contender at present being a palladium-nickel alloy. This performs best, however, if coated with a very thin (~0.1μm) soft gold overlay.[29]

(viii) Silver and its alloys

Silver and its alloys are electrodeposited from baths containing silver and potassium cyanides. Most metal substrates immersed in such a solution would acquire a displacement coating of silver, which is porous and poorly adherent. Hence, a preliminary strike plate is applied from a solution that is high in cyanide and low in silver; this also helps to achieve full coverage.

Pure silver coatings have a hardness of 70–90 HV. They cannot be used in air above about 200°C, oxygen diffusing rapidly through the coating and causing blistering above this temperature. If the substrate contains copper and a nickel undercoat is not used the maximum operating temperature is reduced to 150°C because of interdiffusion.

The hardness can be increased to 130–150 HV by alloying with traces of antimony, selenium or sulphur or with about 3% rhenium. Pure silver has a coefficient of sliding friction against steel of 0.33, but that of electrodeposited Ag-Re alloys is much lower, viz 0.16 at room temperature, 0.18 at 316°C and 0.08 at 621°C.[30]

Silver and silver alloy coatings, because of their high electrical conductivity and corrosion resistance, are widely used in electrical contact applications, their good resistance to sliding wear (Fig. 6.1) making them suitable for use in rotary switches. They are not generally suitable for microelectronics, however, where contact pressures and voltages are low, because of the readiness with which they form tarnish films in sulphur-containing atmospheres.

Silver and silver-lead alloy deposits have been used very successfully in steel-backed plain bearings, but metal cost discourages the growth of this application. They are sometimes used as an anti-galling finish for stainless steel or titanium bolts. Electrodeposited Ag-Re alloy is not only itself a good dry lubricant, but also has the ability to retain molybdenum disulphide grease; it has thus been employed to satisfy particularly arduous and critical lubrication requirements in aircraft.[30]

(ix) Lead and its alloys

The most usual processes employ baths containing metal fluoborates plus free fluoboric acid, some additional boric acid and an additive, such as gelatine, peptone or resorcinol. The additive is essential to prevent roughness and achieve a fine-grained deposit and, in the case of lead-tin plating, to ensure that the deposit contains the minor constituent. Although Pb-Sn and Pb-Sn-Cu are produced directly by codeposition, Pb-In is generally produced by first depositing lead from the fluoborate bath, then applying a layer of indium from a sulphamate bath, and heat-treating to obtain the alloy by interdiffusion.

Lead alloy overlays are applied to the linings of plain bearings that are used in most types of engine. If the lining is aluminium-tin or aluminium-silicon, an undercoat of 2.5–5 μm of nickel is essential; a similar undercoat is sometimes applied if the lining is copper-lead to block tin diffusion from the lead-tin overlay. The function of the overlay is to improve resistance to corrosion, accommodate minor misalignments and embed foreign particles while resisting galling and adhesive wear if the oil film is temporarily disrupted. The thickness required depends on the type of engine—15–30 μm is sufficient for petrol and high-speed diesels, 30–50 μm for the larger, medium- or low-speed diesels.

The as-plated hardness values of the commonly used overlays are: Pb-10% Sn, 8-10 HV; Pb-10% Sn-2%Cu, 12–15 HV; Pb-8%In, 8–10 HV. During service at temperature of 90–170°C, the hardness falls as a result of diffusion. The relative rates of wear of the different overlays, as determined using a dynamically loaded bearing test rig, are shown in Fig. 6.8.

(x) Electrodeposited metal/powder composites

(a) Plating process

The properties of electrodeposits can be modified by the incorporation of solid particles, which are added to the bath and kept in suspension by constant agitation. The idea was first exploited in the production of rough cutting or grinding surfaces by inclusion of coarse diamond or ceramic particles. It later became the basis of a standard method of producing satin nickel finishes without the use of abrasive polishing. Wear resistance applications involve the incorporation of hard particles of moderate size (0.5–5 μm) principally in nickel or cobalt electrodeposits. The dispersoids used commercially are mainly silicon and chromium carbides and alumina, but many other hard, inert materials have been investigated. These coatings have been developed in the UK by BAJ-Vickers and are marketed under the name Tribomet.

Mechanical stirring with an impellor or air agitation can be used to keep the powders in suspension but they are not easily controlled to give good distribution. Pumping solution from the top of the tank and reintroducing it via a cone-shaped base, while sucking in air en route to intensify the mixing effect, is more effective. A plate-pumping device, comprising a horizontal perforated plate maintained in continuous vertical oscillation at the bottom of the tank is also used. Whichever method is adopted, it is not possible to achieve uniform particle incorporation on shapes with blind holes or re-entrant angles.

It is difficult to incorporate powders into chromium deposits. Very vigorous hydrogen evolution is one obstacle; another is the adsorption from chromic acid solution of anions rather than cations on the suspended particles, reducing their chances of being held long enough on the cathode to be engulfed. A recent publication reviews earlier claims and reports the successful incorporation of tungsten carbide in

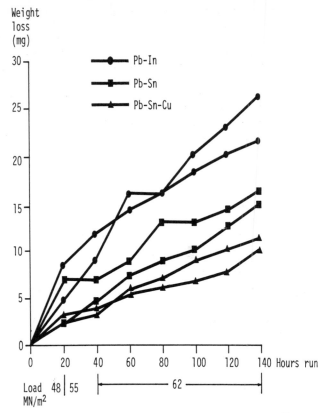

Fig. 6.8 Rates of wear of three types of bearing overlay.[31]

chromium from a standard hexavalent bath; the percentage incorporated was, however, less than 4%.[32] A much larger percentage of alumina has been incorporated into chromium from a trivalent bath, but the problems of depositing a thick, sound deposit from such a solution have yet to be overcome.[33]

Solid lubricants such as molybdenum disulphide can also be incorporated into electrodeposits but the desired self-lubricating properties have not been achieved. Better results are obtained with low friction polymers such as PTFE, as a lower volume percentage is sufficient and has less effect on the cohesion of the coating (see section on electroless nickel/PTFE composites p 97). On a laboratory scale, zinc-graphite composites have been prepared and shown to compare with cadmium in their resistance to corrosion and to galling.[40]

(b) Properties and applications

A nickel-silicon carbide composite was developed initially for coating the rotor and combustion chamber of the Wankel engine and is now used in various small internal combustion engines.[41] Nickel alumina has similar applications.

A cobalt/chromium carbide composite was developed specifically for use on some aero engine components which, because of their geometry, could not be plasma sprayed. Its main use is in the roots of compressor blades where it reduces fretting between the blade root and the disc.[35] At elevated temperatures the coating forms a glaze produced by the oxidation of cobalt, carbide particles supporting the load and anchoring the glaze.[36] On heating for 24 h at 1000°C the carbide changes in composition to M_7C_3 with the result that the volume fraction approximately doubles and the hardness increases from 465 to 600 HV.

The performance of $Co-Cr_3C_2$ composites in sliding and hammer wear tests at 600°C is compared with various other, mainly sprayed, coatings in Fig. 3.10. This shows that the electroplated coating is similar in wear performance to a sprayed Stellite alloy.

Nickel-2% alumina has a hardness of 430 HV compared with 240 HV for the nickel itself; these values fall to 270 and 128 HV respectively, after heating at 650°C.[37] The rate of abrasive wear at room temperature of nickel composites containing 5–9% alumina is less than half that of nickel alone.[38] These composites, because of the relative softness of the matrix, do not resist abrasive wear as well as adhesive wear.

4. Brush plating

Brush plating is the term most commonly used to describe the process of selective electroplating without immersion of the workpiece. A graphite tool is used, wrapped in cotton or polyester material which is soaked in plating solution. This is connected to the positive terminal of a dc supply and the article to be plated connected to the negative terminal. The tool is moved to and fro while plating proceeds; sometimes it is held steady, or moved from side to side, while the work rotates. Special solutions have been developed for brush plating, many contain high concentrations of organically complexed metal ions and some are non-aqueous. Over a hundred different metals and alloys can be deposited in this way.

In the engineering field, brush plating finds its main application in the emergency repair of worn or damaged components. Hard coatings, such as chromium or nickel, may be deposited for wear resistance, or soft coatings, such as lead-tin or silver, for bearing surfaces. The properties of the deposits are, however, not always the same as those from conventional solution; brush-plated chromium, for example, is much softer.

The process is especially useful for treating components that are too large to immerse in a plating bath or belong to mechanisms that are expensive to dismantle completely (eg in ships, and aircraft). Where appropriate, it can be used in mass production:

Table 6.7   Trials of hot forging dies brush plated with Co-Mo alloy[41]

Die type	Sets plated	Results—improvements in die life compared with unplated specimen
Gear selector lever	1	120%
Hand-tool die	1	34%
Coupling-flange die	3	20%, 26% and 21%
Pipe T-piece	1	Production ceased, but estimated it would have been 25%
Link-pin die	1	Production ceased, but estimated it would have been 25%
Gear-blank dies (nitrided)	1	18%, also reduction of sticking of the workpiece to die
Slack adjuster	1	No improvement, but failure normally from cracking not erosion
Rocker-arm insert	1	13%
Large universal-joint dies	2	One pair 77%—the other average life. (The latter was Tufftrided, giving poor electrodeposit)
Large universal-joint dies	4	All four sets produced less than nitrided dies used as a control
Bolster-chisel dies	1	32%
Open-ended spanner die	1	Average life
Turbine-blade—pre-form	1	The plated dies produced 80–250 pieces, but normal life was as low as twenty-five pieces
Turbine-blade dies	1	2200 forged. This is normal life but die will be used again as still in tolerance
Control-linkage die	1	Plated die ran well, but suddenly failed—possibly due to forging cold bar stock
Turbine blade die	1	Used to forge Nimonic alloys had 'normal' life. On removal from forge found to be in tolerance and still usable.
Connecting link rod	1	32%. Improved metal flow and lower forging temperature noted
Suspension-cup die	1	56%. Brush plated over welded area of die—no problems encountered
Ford Transit front-axle dies	2	Improved life but insufficient to be economic
Turbine blade	1	Previous maximum life doubled
Heading die	1	Comparison of pressings from plated and unplated tooling shows effect of reduced die wear using a plated die
Extrusion inserts 0.5 in—2.5 in dia	19	Used for extruding titanium turbine blading. Glass lubricant used. Life significantly increased, but greatly improved surface finish and reduced scrap rate.

at least one computer-controlled brush plating lathe has been made for this purpose.

Reference was made previously to the value of Co-W and Co-Mo as coatings for hot forging dies. As many of the dies used in practice are too large for convenient handling in plating baths, processes were developed enabling the alloy coatings to be deposited under brush plating conditions, and many die sets were successfully coated for industrial trials.[40, 41] The results of these trials carried out on dies brush plated with Co-Mo alloy are summarised in Table 6.7. Although in general the coatings are beneficial the process has yet to become established in production.

5. Electroplating from fused salts

The use of fused salt electrolytes enables metals to be deposited which cannot be plated from aqueous solutions because their deposition potential is much more negative than that of hydrogen. Most of the electrolytes are based on chlorides or fluorides, melting in the range 350–750°C. Diffusion usually occurs at the interface so that coating adhesion is good; at the higher temperatures some surface alloying occurs (metalliding).[42] Elements such as boron, silicon, beryllium, titanium and chromium are by this means incorporated into the surface layers of various substrates, including steel, molybdenum, cobalt, copper and aluminium.

Boride coatings produced on steel in this manner contain both FeB and Fe_2B and penetrate like fingers into the steel. They range up to 100 μm thick and have hardnesses of 1500–2500 HV. Boride coatings with hardnesses greater than 4000 HV have been produced on molybdenum. Beryllide coatings up to 250 μm thick have been produced on copper, titanium, nickel, cobalt and iron. Those on titanium have a hardness around 1500 HV and a much lower coefficient of friction than the metal itself; they resist oxidation on heating in air up to 1000°C. These coatings have exceptional wear resistance, but are expensive and are currently only applied in small scale operations. Some use is made of them in the aerospace industry in both the USA and Europe.[43]

6. Hard anodising

If aluminium or one of its alloys is made anodic in boric acid solution, a very thin layer of oxide is formed on it, the high electrical resistance of the layer preventing its further growth. In sulphuric, oxalic or chromic acid solutions, the layer can however grow to a substantial thickness because the oxide has a finite solubility in these electrolytes and so develops a porous structure, permitting current to pass and more oxide to form. Under ordinary anodising conditions, about half of the aluminium converted to oxide is so dissolved. The material of the film is predominantly alumina, which, although hard, has a high porosity which has an adverse effect on its wear properties.

Hence, where wear resistance is a specific requirement, hard anodising is specified. This means that conditions must be chosen to minimise the solvent action, viz suitable electrolyte, low acid concentration, low temperature, vigorous agitation to avoid temperature rise at the film surface and high current density to minimise process time. In practice a sulphuric acid solution is usually employed, and typical conditions are: concentration, 10% by volume; temperature, 0 ± 5°C; current density, 4.6–23.2 A/dm^2.[44] Sometimes an alternating current component is imposed on the direct current supply, as this allows a further increase in the growth rate of the film. A recent review has given a summary of commercial processes for hard anodising.[45] Ordinary anodising is followed by sealing in boiling water, which hydrates the film, plugs the pores and improves the corrosion resistance. It tends, however, to reduce abrasion resistance by about 20%, so hard anodised parts are not water sealed if maximum abrasion resistance is required.

In conventional anodising, thicknesses vary from 1 to 25μm, depending on the degree of corrosion resistance required while the thickness of hard anodised films are in the range, 25 to 150 μm. Their surface is often rough, so they are commonly ground or lapped to produce a finish of 0.05–0.08 μm CLA.

The hardnesses of ordinary and hard anodised films are within the ranges 200–300 HV and 350–650 HV, respectively. In abrasive jet tests (performed in accordance with BS.1615:1972), the weight of silicon carbide abrasive required to penetrate 1 μm film thickness is 2 and 5 g, respectively.[45] The respective Taber Wear Indexes are 4.3–7.0 and 0.15–0.37 μm per 1000 cycles.[46] Comparative wear data for anodised aluminium and other finishes are shown in Figs. 6.1, 6.2 and 6.3. Hard anodising lowers the fatigue strength of strong wrought aluminium alloys by up to 50%, but dichromate sealing reduces this loss. Hard anodic coatings are widely used on gears, valves, slides, pistons, oil pump and pneumatic components. In the aerospace industry, they are used on helicopter blades, undercarriage parts, brake mechanisms and guidance units for missiles. They are also used on injection moulds for plastics where the operating temperature may be up to 300°C. However, in most of the established uses of anodised aluminium the temperature is lower than this.

References

[1] BUDINSKI, K G. *Selection and Use of Wear Tests for Coatings*. ASTM STP 769, ed Bayer, R G. p 118, 1983

[2] MORISSET, P. *Chromage Dur et Decoratif*. Centre d'Information du Chrome Dur, Paris, 1961.

[3] WELLWORTHY LTD. British Patent 1,320,902, 1973.

[4] GRAHAM, A H and GIBBS, T W. *Properties of Electrodeposits, Their Measurement and Significance*. Chapt. 16, p 255. Electrochem Soc, Princeton, NJ, 1975.

[5] PARKER, K. Hardness and wear resistance tests of electroless nickel deposits, *Plating*, Vol 61, (9), p 834, 1974.

[6] TRSEK, A. The use of thermal sprayed coatings for extended life of aircraft jet engine parts. *Proc. 8th Int. Thermal Spraying Conf.*, Miami, 1, 1976.

[7] WILLIAMS, C and HAMMOND, R A F. The change of fatigue limit on chromium or nickel plating with particular reference to the strength of the steel base. *Trans. Inst. Metal Finishing*, Vol 34, p 317, 1957.

[8] ECKERSLEY, J S. Controlled shot peening strengthens electro and flame deposited coatings. *Plating*, Vol 60, (3), p 214, 1973.

[9] MORGAN, C J and MAYHEW, P R. The effect of shot peening on the fatigue strength of chromium plated titanium alloys IMI 314 and IMI 680, *Trans Inst Met Finishing*, Vol 50, (4), p 141 1972.

[10] *Specification for Electroplated Coatings of Chromium for Engineering Purposes*. British Standards Institution, London, BS 4641:1970.

[11] O'SULLIVAN, D A., Trivalent chromium is basis of plating process, *Chem. Eng. News*, Vol 53, (25), p 16, 1975.

[12] SMART, D, SUCH, T E and WAKE, S J. A novel trivalent chromium electroplating bath. *Trans. Inst. Metal Finishing*, Vol 61, (3), p 105, 1983.

[13] Information from Permalite Chemicals Ltd, Ridgeway, Iver, Bucks. SL0 9JJ.

[14] LONGO, F M. Plasma and flame sprayed coatings satisfy hard chromium plate applications. *Proc. 8th Int. Thermal Spraying Conf.*, Miami, p 342, 1976.

[15] DIBARI, G A. Nickel plating. p 278, *Metal Finishing Guidebook and Directory*. Metals and Plastics Publications Inc, Hackensack NJ, 1981.

[16] DILL, A J. Sulphur-free organic hardening agents for electrodeposited nickel, *Plating and Surface Finishing*, Vol 62, (8), p 770, 1975.

[17] WEARMOUTH, W R and BELT, K C. The mechanical properties and electroforming applications of nickel-cobalt electrodeposits, *Trans. Inst. Metal Finishing*, Vol 52, 3, p 114, 1974.

[18] STILL, F A and DENNIS, J K. The use of electrodeposited cobalt alloy coatings to improve the life of hot forging dies, *Electroplating and Metal Finishing*, Vol 27, 9, p 9, 1974.

[19] GRECO, V P, BALDAUF, W and COTE, P J. Electrodeposition of cobalt-rhenium alloys (evaluation of electrolysis and deposit properties), *Plating*, Vol 61, (5), p 423, 1974.

[20] SAFRANEK, W H. *The Properties of Electrodeposited Metals and Alloys—A Handbook*. Elsevier NY, 1974.

[21] WILSON, F G and WARD, R. The dry wear behaviour of porous cobalt, *Cobalt*, No 55, p 87, 1972.

[22] BUCKLEY, D H. Adhesion, friction and wear of cobalt and cobalt-base alloys, *Cobalt*, No 38, p 20, 1968.

[23] DENNIS, J K and STILL, F A. The use of electrodeposited cobalt alloy coatings to enhance the wear resistance of hot forging dies, *Cobalt*, p 17, 1975(1).

[24] SCOTT, D and McCULLOCH, P J. Hardenable electrodeposited coatings for rolling bearings. *Electrodeposition and Surface Treatment*, 1, p 21, 1972.

[25] KLINGENMAIER, D J. Hard iron plating of aluminium pistons. *Plating*, Vol 61, (8), p 741, 1974.

[26] LAI, S H F and McGEOUGH, J A. Electroforming of iron components, *Trans. Inst. Metal Finishing*, Vol 57, (2), p 70, 1979.

[27] NOBEL, F I, THOMPSON, D W and LEIBEL, J M. An evaluation of 18 Karat and 24 Karat hard gold deposits for contact applications. *Plating*, Vol 60, (7), p 720, 1973.

[28] ANTLER, M. Tribology of metal coatings for electrical contacts. *Thin Solid Films*, Vol 84, p 245, 1981.

[29] GRAHAM, A H, PIKE-BIEGUNSKI, M J and UPDEGRAFF, S W. Evaluation of palladium substitutes for gold, *Plating and Surface Finishing*, Vol 70, (11), p 52, 1983.

[30] TURNS, E W. Silver-rhenium alloy plating: industrial development, *Plating*, Vol 50, (2), p 127, 1971.

[31] EASTHAM, D R and CROOKS, C S. Plating for bearing applications. *Trans. Inst. Metal Finishing*, Vol 60, (1), p 9, 1982.

[32] NARAYAN, R and SINGH, S. Composite chromium coatings containing tungsten carbide, *Metal Finishing*, Vol 81, p 45, March 1983.

[33] ADDISON, C A and KEWARD, E C. The development of chromium based electrodeposited composite coatings. *Trans. Inst. Metal Finishing*, Vol 55, (2), p 41, 1977.

[34] DONAKOWSKI, W A and MORGAN, J R. Zinc/graphite—a potential substitute for anti-galling cadmium. *Plating and Surface Finishing*, Vol 70, (11), p 48, 1983.

[35] KEDWARD, E C, ADDISON, C A and TENNETT, A A B. The development of a wear resistant electrodeposited composite coating for use on aero engines. *Trans. Inst. Metal Finishing*, Vol 54, (1), p 8, 1976.

[36] CAMERON, B P, FOSTER, J and CAREW, J A. The effect of carbide content and pre-heat treatment on the oxidation characteristics of cobalt-chromium carbide electrodeposits in air at 1000°C. *Trans. Inst. Metal Finishing*, Vol 57, (3), p 113, 1979.

[37] SINHA, P K, DHANANJAYAN, N and CHAKRABATI, H K. Electrodeposited nickel-alumina composites. *Plating*, Vol 60, (1), p 55, 1973.

[38] BRONSZEIT, E, HEINKE, G and WIEGAND, H. Uber die mechanischen Eigenschaften galvanisch hergestellter Nickel dispersionsschichten mit Einlagerungen von Al_2O_3 und SiC. *Metall*, Vol 25, p 470, 1971.

[39] KEDWARD, E C. Electrodeposited composite coatings, *Cobalt*, p 53, 1973(3).

[40] LODGE, K J, STILL, F A, DENNIS, J K and JONES, D. The application of brush plated cobalt alloy coatings to hot- and cold-work dies. *J. Mech. Working Technology*, Vol 3, (63), 1979.

[41] *Dies and Moulds—Research on the Problems involved in Manufacture*. Science and Engineering Research Council, 1982.

[42] COOK, N C, Metalliding. *Scientific American*, Vol 221, (2), p 38, 1969.

[43] WITHERS, J C, PERRY, J E and FOSNOCHT, B A. Chapt 5, Pt 1, Vol VII, *Techniques of Metals Research*, ed Bunshah, R F. Interscience Publishers, New York, 1972.

[44] THOMPSON, D A. Production of hard anodic coatings. *Trans. Inst. Metal Finishing*, Vol 54, (2), p 97, 1976.

[45] KAPE, J M. Hard anodising. *Inst. Metal Finishing, Conference Preprints*, 1983.

[46] THOMAS, R W. Measurement of hardness, wear index and abrasion resistance of anodic coatings on aluminium. *Trans. Inst. Metal Finishing*, Vol 59, (3), p 97, 1981.

Additional reading

Modern Electroplating. ed Lowenheim F A. John Wiley & Sons Inc. NY, 3rd edn, 1973.

CHAPTER 7

Chemical treatments

1. Introduction

Chemical treatments here refer to processes in which an article immersed in an aqueous solution is coated without the application of an applied electromotive force. There are many such treatments including chromate passivation, which confers improved corrosion resistance, and a range of metal colouring procedures. Some of the latter produce thin films, such as the black oxide coatings on steel which, when impregnated with lubricant, have a minor effect on frictional properties. The only important chemical treatments, however, in relation to wear resistance, are electroless nickel plating and phosphating. A composite coating based on chromium oxide has been developed comparatively recently and several areas of application are being investigated.

2. Electroless plating

Electroless plating is the deposition of metals by controlled autocatalytic chemical reduction. Metal ions in solution are reduced to the metal by a chemical agent, the surface of the workpiece catalysing the reaction so that a coating is produced upon it.

A major advantage of electroless deposition is its ability to coat articles of complex shape with a layer of almost uniform thickness. In addition, it requires no power supply or electrical contacts, and it can be applied to non-conductors with suitable pre-treatment. Because of the high cost and relative inefficiency of the reducing agents, and the solution losses that are unavoidable with this type of process, electroless plating is more expensive than electroplating.

Electroless copper is important in plating plastics and is essential in the through-hole plating of printed circuit boards, but has little significance in connection with wear resistance. Electroless silver may have a role to play; coatings applied on top of chromium or electroless nickel improve wear properties.[1] Processes have also been described for cobalt, gold, platinum and various alloys, but no major applications have been reported. The main industrial use of electroless plating is the deposition of nickel coatings containing phosphorus or boron.

(i) Coating process

(a) Plant and equipment

The equipment required is broadly similar to that used in electroplating. Solutions are contained in polypropylene tanks, heated by immersion heaters or hot-water coils and filtered continuously. Pumps must withstand temperatures up to 100°C and have no metal parts in contact with the solution. Some industrial baths are continuously regenerated, nickel salts and reducing agent being added to solution that has been withdrawn from the reaction vessel. Usually, however, additions are made directly to the process tank. The solution is agitated by air or reciprocating movement of the workpieces. Small parts can be economically plated in barrels.

(b) Surface preparation

Similar preparatory procedures to those described in Chapter 6 for electroplating are employed. Only catalytic metals, however, plate spontaneously on immersion; these include iron, nickel, cobalt, gold, silver and platinum group metals. Non-catalytic metals, such as copper need to be started by touching with a steel article or brief application of a cathodic potential. Other non-catalytic metals are kept out of contact with the solution as they poison it. It is therefore recommended that basis metals containing more than trace amounts of antimony, arsenic, bismuth, cadmium, lead, magnesium, tin or zinc (except bronzes and brasses) have an undercoat applied to them of a few microns of electroplated copper or nickel. A similar undercoat may be applied to basis materials containing chromium, molybdenum and titanium to achieve good adhesion.

Electroless nickel can be applied to non-metallic materials, such as glass, ceramics or plastics.[2] Plastics are first etched, then the surface is sensitised by a dip in acidified stannous chloride solution and activated by brief immersion in a palladium chloride solution, before rinsing and placing in the plating bath.

(a) Plating

Most electroless nickel is deposited using a solution of sodium hypophosphite ($NaH_2PO_2.H_2O$) as the reducing agent. The bath contains only a low concentration of nickel ions (5–6 g/l) usually added as sulphate, but present in the solution as a complex with carboxylate ions. Catalyst poisons such as sulphur-compounds (thiourea), oxyanions (MnO_4^-) and heavy metals (Pb) are usually added to provide the necessary stability. Other additions may be made to improve the brightness of the deposits. The pH is controlled within the range 4.0–5.5 and temperatures in the range 80–95°C. Deposition rates can be as high as 20 μm/h but the average over the life of a bath is usually considerably less. The solution is regenerated by adding fresh reagents, but, because of by-products,

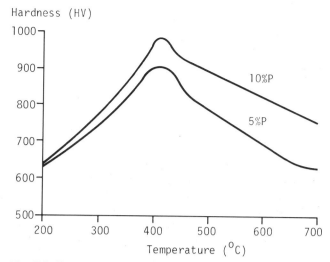

Fig. 7.1 Room temperature hardness of Ni-P as a function of heat-treatment temperature (1h at temperature, reached at 20°C/min.)

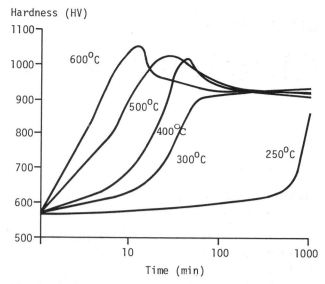

Fig. 7.2 Room temperature hardness of Ni-P as a function of time at heat treatment temperature.

has to be discarded after a time in which between five and ten times the initial nickel content may have been deposited. The coatings contain between 2 and 15% phosphorus.

In other processes, compounds of boron instead of phosphorus are used as reducing agents. N-dimethyl-amine borane and N-diethylamine borane are effective over a wide range of conditions and give coatings with low boron contents (0.2–4.0%). These baths can be operated at relatively low temperatures, and are used extensively in the electronics industry. Sodium borohydride (NaBH₄) is the reagent mainly used for engineering purposes. To prevent the formation of nickel boride, the baths are operated at a pH above 12 and a strong complexing agent, such as ethylene diamine, is present to avoid the precipitation of nickel hydroxide. A stabiliser, usually thallium sulphate, is also added. The operating temperature is in the range 90–95°C, and the plating rate is 12–20 μm/h. The deposits typically contain 5% boron and as much as 1% thallium.[3]

It is possible to incorporate fine particles of, eg silicon carbide, diamond, alumina and PTFE in electroless nickel deposits. In this process powerful stabilisers are added to the bath to prevent the particles acting as nuclei for metal deposition. Relative motion of solution and substrate must be controlled to obtain satisfactory particle distribution.

(d) Post-plating treatments

Heat treatment may be carried out to reduce hydrogen embrittlement, improve adhesion on some substrates, or increase hardness. The temperature recommended to reduce hydrogen embrittlement is 190–220°C, for a time that increases with the tensile strength of the steel substrate. The temperature and time of treatment required to improve adhesion depends on the nature of the basis metal and varies between 130°C for 1–1.5h for age-hardened aluminium alloys and 280°C for 10h for titanium alloys.

The temperature and times of treatment appropriate for increasing deposit hardness depend on composition and the increase in hardness required. Typically, maximum hardness is attained after heating at 400°C for 1h. Rapid heating and cooling should be avoided. An inert or reducing furnace atmosphere is preferred but hydrogen should not be used for steel parts of high tensile strength. It should be borne in mind that heat treating for maximum hardness produces a contraction of the coating, resulting in microcracking, which adversely affects the impact strength and lowers the protection against corrosion.

(ii) Properties

(a) Nickel-phosphorus

Alloys containing 7–13% phosphorus are the most suitable for engineering applications. As deposited, they comprise a supersaturated solid solution of phosphorus in a very fine grained nickel matrix with a preferred orientation of <111> planes to the surface.[4] Striations parallel to the substrate are revealed by etching, and probably correspond to fluctuations in phosphorus content. The surface is lustrous and may be fully bright, but there is no levelling of a rough substrate. The hardness is within the range 450–650 HV and most commonly around 500 HV. Some friction coefficients, measured with and without lubrication, are reported in Table 7.1.

Table 7.1 Friction coefficients of electroless nickel coatings[5]

Couple	Coefficient of sliding friction	
	Dry	Lubricated
Electroless Ni/Ni	seizure	0.26
Electroless Ni/electroless Ni	0.45	0.25
Electroless Ni/Cr	0.43	0.30
Electroless Ni/steel	0.38	0.21
Electroless Ni/cast iron	0.16	0.08
Cr/Cr	0.43	0.26
Cr/steel	0.21	0.15
Steel/steel	seizure	0.20

Table 7.2 Taber wear test results for Ni-P coatings[6]

Coating	Heat treatment, °C/h	Taber Wear Index
Ni-P	—	18
Ni-P	288/10	6
Ni-P	400/1	6
Ni-P	540–650/1	4
Cr	—	3

When electroless nickel is heated above 200°C, nickel phosphide (Ni_3P) precipitates, with a resulting increase in room temperature hardness. The hardness developed depends on time, temperature and composition, as indicated in Figs. 7.1 and 7.2., which are based on data from various sources. Typical hardnesses measured at temperature are: 400°C, 420 HV; 485°C, 230 HV; 525°C, 160 HV. The effect of heat treatment on weight loss in Taber tests performed

Table 7.3 Falex tests, electroless nickel and electrodeposited chromium[7]
(Test conditions and key to symbols below)

V-blocks				Rotating pins			
Steel	Coating	Treatment	Wear, mg	Steel	Coating	Treatment	Wear, mg
A	—	—	61.5	—	Cr	—	0.2
A	—	—	2.3	—	Ni-P	Q	0.1
A	—	—	7.0	—	Ni-P	R	0.6
A	—	—	6.8	—	Ni-P	S	0.3
A	—	—	2.2	—	Ni-P	T	0.1
B	—	—	0.5	—	Ni-P	P	17.5
B	—	—	0.3	—	Ni-P	Q	3.9
B	—	—	0.7	—	Ni-P	R	3.7
B	—	—	1.5	—	Ni-P	S	3.0
B	—	—	0.8	—	Ni-P	T	5.6
B	—	—	0.2	—	Ni-P	U	6.4
B	—	—	15.0	—	Cr	—	6.2
C	—	—	0.7	—	Ni-P	R	1.5
—	Cr	—	0.5	B	—	—	1.9
—	Ni-P	—	6.6	B	—	—	0.2
—	Ni-P	Q	1.2	B	—	—	0.1
—	Ni-P	R	0.4	B	—	—	0.1
—	Ni-P	S	0.5	B	—	—	0.2
—	Ni-P	T	1.4	B	—	—	0.1
—	Ni-P	Q	3.9	D	—	—	galling
—	Ni-P	P	20.8	D	—	—	galling
—	Ni-P	Q	1.4	—	Ni-P	Q	3.5
—	Ni-P	R	1.0	—	Ni-P	R	1.7
—	Ni-P	S	0.5	—	Ni-P	S	2.0
—	Ni-P	T	0.6	—	Ni-P	T	4.6
—	Ni-P	U	4.1	—	Ni-P	U	1.4
—	Ni-P	T	0.4	—	Ni-P	R	0.1
—	Ni-P	Q	2.0	—	Cr	—	0.5
—	Ni-P	R	17.0	—	Cr	—	0.5
—	Cr	Q	galling	—	Cr	—	galling
—	Cr	—	0.6	—	Ni-P	R	galling

Conditions and Key to Table 7.3

Test conditions
   load: 22.5 kg for 5 min, 90 kg for 60 min then 180 kg for 40 min
   lubricant: white mineral oil
   hardness of electrodeposited chromium: 1100 HV

Key to steels:

Symbol	SAE No	Hardness	
		HRC	HRB
A	1137	20–24	
B	9310	60	
C	4140	60	
D	3135		90

Key to heat treatment and hardness of electroless nickel
(phosphorus content: 9–10%, hardness without heat treatment: 590 HV)

Symbol	Temperature, °C	Time, h	Hardness HV
P	200	2	640
Q	288	2	880
R	288	16	1050
S	400	1	1100
T	540	1	750
U	600	1	695

under 1 kg load is shown in Table 7.2.

Table 7.3. contains the results of Falex wear tests undertaken with the rotating pins, or the V-blocks, or both, plated with electroless nickel and submitted to various heat treatments. Some tests were done on chromium coatings for comparison. At best, electroless nickel wears more slowly than chromium and produces much less wear of the counterface, if it is hard; the softer steel pins, however, gall in contact with electroless nickel plated blocks. With both faces coated, heat-treated electroless nickel performs well against a similar coating, chromium plated pins wear slowly against nickel plated blocks but chromium against chromium fails. The performance relative to chromium in these tests is not matched in Fig. 6.1., which refers to block-on-ring tests conducted at low load but without lubricant. Some further block-on-ring tests of electroless nickel, with lubrication, are reported later in Table 7.7.

Electroless nickel is a useful coating for titanium and aluminium alloys, which are highly susceptible to galling. The seizure load in a Falex test for a titanium shaft rotating in steel V-blocks is 4 MN/m^2 but this rises to 22.8 MN/m^2 if the shaft is coated with electroless nickel.[8]

As-deposited electroless nickel coatings, provided they are at least 25 μm thick, improve the fretting wear resistance of quenched and tempered steels; 10 μm coatings are beneficial if they have been heat treated to a hardness above 700 HV.[9] The fatigue strength of high tensile steels is lowered by electroless nickel plating, but prior shot peening moderates the loss.

Using ring shear tests, adhesion values for electroless nickel on a variety of substrates have been measured up to the following maxima:

aluminium alloys	250 MN/m^2
beryllium copper	410 MN/m^2
4340 steel	470 MN/m^2
HP 9–4–20 steel	520 MN/m^2

To achieve the last result, it was necessary to etch anodically in sulphuric acid and strike in an acid nickel solution before plating with electroless nickel.[10]

Electroless nickel-phosphorus alloys are resistant to corrosion in many environments.[6] For maximum resistance, however, they require a relatively high phosphorus content and should preferably be deposited from solutions that are low in sulphur-containing and heavy metal stabilisers. Heat treatment at around 400°C to achieve maximum hardness may cause cracking and so lower the degree of corrosion protection. Heat treatment at 650°C has, however, been used to increase resistance to marine environments.

(b) Nickel boron

Coatings of electroless Ni-B alloy are grey in colour and lacking in lustre. They appear to be partly amorphous and partly microcrystalline, with a crystal size in the range 15–60 Å. The as-deposited hardness in the range 650–750 HV is substantially higher than that of Ni-P deposits. Heat treatment at 400°C for 1 hr raises this to approximately 1200 HV, owing to

Table 7.4 Taber wear test results for Ni-P and Ni-B coatings[3]

Coating	Heat treatment, °C/h	Taber Wear Index
Electrodeposited Ni	—	25
Ni-10.5% P	—	19
Ni-10.5% P	400/1	10
Ni-5% B	—	9
Ni-5% B	400/1	3
Cr	—	3

Table 7.5 Yarnline wear resistance of electroless nickel alloys[11]

Alloy	Condition	Hardness, HV	Wear rate, 10^{-8} cm^3/h
Ni-10% P	as deposited	550	648
Ni-10% P	400°C for 1 h	950	65.5
Ni-10% P	600°C for 1 h	600	52.7
Ni-10% P	750°C for 1 h	550	55.1
Ni-4% B	as deposited	700	216
Ni-4% B	425°C for 1 h	1250	39.8
Ni-4% B	600°C for 1 h	1100	26.0
Test standard:			
Vasco 7152 tool steel		520	5.4

the precipitation of nickel borides (principally Ni_3B). Long term treatments of many weeks' duration at temperatures of 200–300°C can produce hardnesses of 1700–2000 HV.[3]

Some comparative Taber test results are given in Table 7.4, indicating that the wear resistance of heat treated Ni-B coatings is as good as that of chromium under these conditions; the performance of heat-treated Ni-P is inferior to that reported in Table 7.3. Some yarnline wear resistance measurements made under the same conditions as those in Table 6.4 are reported in Table 7.5. The rate of wear is expressed in different units, but reference to the behaviour of the standard tool steel material shows that, while Ni-B is distinctly superior to Ni-P, and heat treatment has a beneficial effect on both alloys, their performance is inferior to that of chromium. The incorporation of hard particles in the coating, however, considerably increases wear resistance (Table 7.8.)

The coefficient of friction of Ni-B versus steel is 0.12–0.13 with, and 0.43–0.44 without, lubrication. Ni-B is as resistant as Ni-P to corrosion by alkalis and solvents and as susceptible to corrosion by strongly oxidising media. Acids and ammonia solutions, however, which are moderately corrosive towards Ni-P, are severely corrosive towards Ni-B.[3]

(c) Nickel/powder composites

Up to 50% by volume of particles in the size range 1–10 μm can be incorporated in electroless nickel, but most use is made of deposits containing 25–30% by volume. A thickness of 25–50 μm is sufficient for many applications although thicker coatings are possible. The most important dispersoids are polycrystalline diamond, silicon carbide, alumina in platelet form and PTFE.[12] Table 7.6 gives the values of the Taber Wear Index determined on a number of composite coatings based on electroless Ni-P containing 15–20% by volume of incorporated particles. Carbides and diamond give the lowest wear (similar

Table 7.6 Taber wear test results for Ni-P/Powder coatings[7]

Powder in Ni-P	Taber Wear Index	
	As deposited	Heat treated
—	18	8
Graphite	15	8
Chromium carbide	8	2
Tungsten carbide	3	2
Aluminium oxide	10	5
Titanium carbide	3	2
Silicon carbide	3	2
Boron carbide	2	1
Diamond	2	2
For comparison:		
hard anodised Al	2	
Cr	3	

to that of hard anodised aluminium or chromium). Heat treatment at 260–290°C brings about a further increase in wear resistance.

In Table 7.7 are listed the results of LFW–1 tests for steel rings (hardness 65 HRC) against plated blocks and plated rings against steel blocks (hardness 60 HRC). The coatings are 25 μm thick, the speed 72 rev/min and the lubricant white oil. The maximum load and duration are 68 kg and 5000 cycles in the tests on plated blocks. Heat treatment considerably reduces the wear of plated blocks. The incorporation of particles of silicon carbide, boron carbide, diamond or tungsten carbide has a similar effect without heat treatment but increases the wear on the steel ring. In the tests on plated rings, which were carried out at a higher load (284 kg) and for a longer time (25,000 cycles) than the tests on plated blocks, severe wear of the non-heat treated coating occurred. This was reduced to a very low level on heat treating for 1 hr at 400°C. Again incorporation of hard particles in general reduced the wear of the non-heat treated coating and increased the wear of the counterface. The anomalously high amount of wear of the plated ring containing tungsten carbide particles is unexplained.

Electroless Ni-B coatings containing hard particles can have a very high resistance to yarnline wear.

Some results from a test in which a slurry of abrasive submicron alumina particles is applied to the moving yarnline before it reaches the specimen, so accelerating wear by a factor of 600 times are shown in Table 7.8. Reference to the wear rate recorded on the test standard shows that the composites (especially those containing diamond particles) have a resistance to wear that is orders of magnitude greater than that of as-plated or heat-treated Ni-P or Ni-B (Table 7.5) or electrodeposited chromium (Table 6.4).

Recently coatings of electroless Ni-P with incorporated PTFE particles have become commercially available. A proprietary coating, Niflor, contains about 25% by volume PTFE and has a hardness of 250 HV as deposited, rising to 400 HV after heating at 300°C for 4 hr (the matrix hardness increasing from 500 to 1000 HV). Other heat treatments are possible that have little effect on the matrix but sinter the PTFE both within the coating and on its surface. The polymer contributes a lubricating effect to engineering parts, which is maintained as the coating wears away. Measurements comparing Niflor with an anodised and dichromate sealed aluminium surface, in reciprocating sliding tests with a stainless steel counterface, have shown that the Ni-P/PTFE coating gives a lower and more consistent coefficient of friction over a much longer sliding distance than the anodised finish.[13]

In Taber tests, the PTFE composite is less wear resistant than electroless Ni-P deposits; it does, however, show increasing resistance as the coating wears and the amount of polymer debris increases.[14]

Table 7.8 Yarnline wear resistance of Ni-B composites in accelerated tests[11]

Coating	Wear rate, μm/h
Ni-0.5%B	23,000
Ni-0.5%B/10% by vol 10μm SiC	278
Ni-0.5%B/10% by vol 8μm Al$_2$O$_3$	109
Ni-0.5%B/10% by vol 9μm natural diamond	10.2
Ni-0.5%B/10% by vol 9μm synthetic diamond	5.1
Test standard:	
Vasco 7152 tool steel	1,450

Table 7.7 LFW–1 tests on electroless nickel composites[7]

Block	Ring	Treatment of coating, °C/h	Coating hardness, HV	Final kinetic friction coefficient	Block loss, mg	Ring loss, mg
Ni-P	steel	—	585	0.128	9.0	0.6
Ni-P	steel	400/1	1064	0.120	2.3	0.5
Ni-P/PTFE	steel	400/1	900	0.113	2.8	0.9
Ni-P/SiC	steel	—	724	0.128	1.0	1.0
Ni-P/B$_4$C	steel	—	724	0.107	2.2	0.8
Ni-P/B$_4$C	steel	lapped	724	0.103	2.0	0.8
Ni-P/WC	steel	—	707	0.133	3.2	3.1
Ni-P/diamond	steel	400/1	1100	0.120	1.2	4.5
Steel	Ni-P	—	625	0.086	0.2	102
Steel	Ni-P	400/1	1060	0.106	0.0	1
Steel	Ni-P/SiC	260/10	1090	0.127	5.1	0.1
Steel	Ni-P/B$_4$C	—	670	0.094	6.1	3
Steel	Ni-P/WC	—	700	0.121	2.9	96

(iii) Applications

The use of electroless nickel has grown steadily over the last few decades because of its ability to cover complex shapes with uniform coatings that have good wear and corrosion resistance in many environments. Such coatings have been applied to many components of cars and aircraft, and have established themselves particularly on carburettor parts. In the oil and chemical industries they are used on valves, pumps, filters and some vessels. They are sometimes applied to components of plastics injection moulding machinery, eg screws, nozzles and moulds. Their use has been reported on textile, printing and hydraulic equipment, and on tools and surgical instruments. The coatings have proved useful on dies for zinc alloy die castings, where their relatively low thermal conductivity prevents premature freezing. They are used in the reclamation of under-sized parts, in thicknesses up to about 125 μm; an electrodeposited nickel undercoat may be used if a greater thickness is required, but machining to size is then likely to be necessary before the electroless layer is applied.

Ni-P is most likely to be used for all the above applications, but in Germany and Japan, Ni-B might be employed in a significant proportion of cases. Because of its higher temperature capability (mp 1080 compared with 890°C) and greater hardness, Ni-B has a special appeal for particular applications, eg as coatings for glass moulding equipment and some cutting tools.

Electroless nickel/powder composites have found a variety of uses; for example, those containing silicon carbide are used on injection screws and moulds for plastics, on metal forming dies and foundry patterns,[15] and those containing diamond on broaches, reamers, yarn guides, valves, clutches and microfinishing tools.[16] Other applications are on tools for working wood or filled resins and equipment for high speed handling of paper.[12] Examples of components that benefit substantially from the use of an electroless nickel/PTFE coating are: stainless steel butterfly valves, moulds for rubber, carburettor parts, aluminium air cylinders, pumps, rotors, mould cores and controlled-torque fasteners.[13]

2. Phosphating

Phosphating is a conversion process in which a metal surface reacts with an aqueous solution of heavy metal primary phosphate plus free phosphoric acid to produce an adherent layer of insoluble complex phosphates. It is used mainly on mild or low alloy steels and, to some extent, on aluminium, zinc and their alloys. The principal phosphates involved are those of iron, zinc and manganese. As indicated in Table 7.9 there are various applications for phosphate coatings, depending on their type and weight per unit area. The biggest application is the pretreatment of mild steel prior to painting, where coatings improve adhesion and corrosion resistance. Coatings can be applied by immersion or alternatively by spraying or brushing.

Thick coatings, as required for electrical insulation or corrosion resistance in combination with grease or wax finishing, are applied by immersion only. This is generally true also of the coatings of intermediate thickness used in engineering applications. These fall into two categories, although both rely on the ability of phosphate coatings to reduce friction and wear. The coatings serving as aids in the cold deformation of metals are generally based on zinc phosphate, while those that improve the lubrication properties of ferrous components are based on manganese phosphate.

(i) Coatings for cold working

(a) Coating process

The usual procedure is to apply a phosphate coating, then partially react it with soap solution to impregnate the surface with an insoluble heavy metal soap, which acts as lubricant. Zinc phosphate is preferred to manganese or iron phosphate because zinc soaps have the best lubricating properties, the crystals of zinc tertiary phosphate deform under stress, the coatings are formed more quickly and the process is more economical.

The metal, which may be in the form of wire or tube for drawing, sheet for pressing or slugs for

Table 7.9 Applications of phosphating[17]

Application	Nature of coating and weight per unit area, g/m²	
Preparation for painting	Amorphous iron phosphates:	0.3–0.9
	Zinc and iron phosphates:	1.5–4
Corrosion resistance when finished with grease or wax	Iron and zinc phosphates:	10–30
	Manganese and iron phosphates:	10–30
Electrical insulation of sheet silicon steel	Magnesium iron and magnesium phosphate	10–30
Cold working: tube and wire drawing, deep drawing, extrusion	Zinc and iron phosphates:	5–20
Improvement of lubrication properties, antiseizure	Manganese, iron and possibly nickel phosphates:	5–15

extrusion, is cleaned, pickled and thoroughly rinsed, then phosphated at about 70–85°C in an acid solution of zinc primary phosphate. After rinsing, the work is immersed in a soap bath, ie a solution of the sodium salt of a long-chain fatty acid, preferably sodium stearate. This produces a layer of zinc soap, 0.1–0.5 g/m^2, firmly bonded to the phosphate coating. Although soaps are more effective lubricants, mineral or vegetable oils can also be used.[18] Alternatively, coils of wire or batches of rod are dipped in borax, lime or sodium metasilicate after phosphating.

The complexity of these process routes, involving six or seven wet-chemical steps, provided the incentive for the development of a combined treatment using mineral oils containing phosphate esters and free phosphoric acid. The films produced have a coating weight of 5–30 g/m^2 but the phosphate layer itself is only 1–6 g/m^2 (ie 5–30% of the total).[19] This type of process has largely supplanted zinc phosphate plus soap for tube drawing in Germany and Italy, but has made slower progress in the UK. It is more convenient in operation and gives a better finish but, being oil-based, is more expensive.

Stainless steels and nickel-chromium alloys are resistant to attack by phosphoric acid and so cannot be phosphated, but an alternative process can be used based on oxalic acid and oxalates. If sodium thiosulphate or metabisulphite is added as accelerator, a heavy and adherent coating, containing both oxalate and sulphide, can be produced forming the basis of a highly effective drawing lubricant.

(b) *Properties and applications*

Zinc phosphate/soap coatings applied by immersion to the inside and outside of steel tubes enable tube drawing to be performed at high speeds, up to 100 m/min., with a reduced frequency of inter-pass pickling and annealing treatments and improved product quality. The single-stage phosphating process performed in an oil medium is capable of giving an even better finish, eg on seamless tube intended for shock absorber manufacture. On tubes with uneven surfaces and raised seam welds, phosphate/soap is generally better.

Mild steel wire can be drawn satisfactorily using stearates in powder form as lubricant. For high tensile steel wire, phosphating followed by a dip in a hot lime suspension is used. Stainless steel or nickel alloy wire and rod, coated with oxalate/sulphide films and treated with soap/borax lubricants, can be drawn at high speed on multi-hole machines.

Zinc phosphate coatings are also used to facilitate the cold heading of bolts and the stamping of components such as car bumpers. They are essential in the cold extrusion of steel components using high speed steel or cemented carbide dies and plungers; reactive lubrication with soap is standard in this application.

(ii) Coatings for moving parts

(a) *Coating process*

The best phosphate coatings for moving parts are

Table 7.10 Results of Shell 4-ball tests[17]

Load, kg	Mean scar diameter, mm			
	Not phosphated	Phosphated		
		Zn A	Zn B	Mn
340				weld in 10s
320				1.28
300				1.26
280			weld in 10s	
260			0.98	
240		weld in 10s	0.88	1.10
220		0.90	0.80	0.95
200		0.80	0.75	1
120	weld			
110	3.25			
100	3.08			0.7

those based on manganese, which are particularly hard. On steel components, some iron is usually present in the coating and nickel may be added to the solution.[18]

After cleaning, the workpiece is immersed in an agitated, aqueous suspension of fine crystals of manganese phosphate, some of which adhere to the surface and are carried into the phosphating bath, where they serve as crystallisation nuclei and promote the formation of a fine, compact coating. Phosphating is carried out at 96—99°C for a period of 9–15 min, and is followed by a rinse in hot water or sometimes a passivating rinse in chromic acid solution.

After drying, the coating is lubricated by one of the following, depending on the subsequent application: (a) mineral oil; (b) molybdenum disulphide in varnish or grease; (c) colloidal graphite; (d) leaded grease or (e) soluble oil.

(b) *Properties and applications*

Homogeneous, adherent and absorbent phosphate coatings are obtained on steels containing not more than about 5% of alloying constituents such as Ni, Cr and Mo. Case hardened surfaces can be treated without any loss of properties. The non-metallic layer prevents metal-to-metal contact and is highly resistant to welding and seizure. Because of its porosity it absorbs lubricant and maintains lubricity under load, even if the supply of lubricant is briefly interrupted.

The coating weight per unit area is in the range 5–15 g/m^2. In practice, this means a thickness between 4 and 8 μm, occasionally rising to 12 μm. If close tolerances are required, the coating may be buffed.

Table 7.10 shows the results of tests made on a Shell 4-ball machine, using 12.7 mm diameter balls of alloy steel tempered to a hardness of 62–65 HRC. The test time was one minute and the lubricant pure mineral oil. The two zinc phosphating baths both contained some nickel, but A was accelerated with chlorate and B with nitrate. It is clear that the maximum weld load is highest in the presence of the manganese phosphate coating. The scar diameter is greater, however, for a given load than is observed with the zinc phosphate coatings—probably because manganese phosphate is harder and therefore more abrasive.

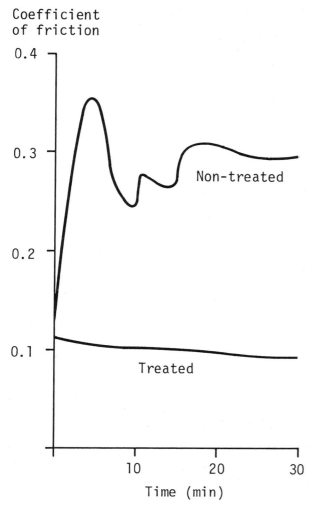

Fig. 7.3 Effect of manganese phosphate coating on coefficient of friction in Amsler test.[17]

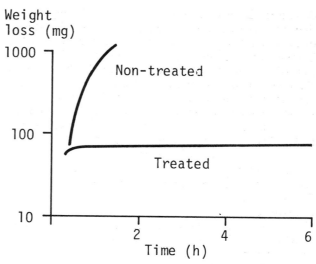

Fig. 7.4 Effect of manganese phosphate coating on weight loss in Amsler test.[17]

Figs. 7.3 and 7.4 show the variations with time of the coefficient of friction and weight loss in pure sliding Amsler tests, using untreated and manganese phosphate coated alloy steel discs. The load was 100 N and the lubricant mineral oil. It is clear that the coating maintains the friction coefficient at a lower and much steadier value, while reducing the rate of wear after a short initial period.

There is some evidence that phosphate coatings may lower the fatigue strength of steels, probably because the etching that occurs during processing creates stress raisers. As with most aqueous processes there is a danger of hydrogen embrittlement. BS.3189:1973 recommends that steels having tensile strengths in the range 1000–1383 MN/m² should be heated for at least 4h at 150–200°C after phosphating.

Manganese phosphate coatings break down at about 300°C in air, but can tolerate higher temperatures in a lubricated engine. They are resistant to corrosion, especially if oiled—an important consideration during storage before assembly.

The principal value of these coatings is in running-in new machinery, but the benefits they secure may survive for long periods. They are widely used in the car industry on differential gears, gear box components, cam shafts and push rods, bearing races, the shafts of oil pumps, rocker arms, pistons and rings, cylinder liners, valves and valve guides and tappets. In the oil industry, they are applied to the taper threaded collars joining drill pipe sections together and connecting the cutting bit to the bottom section; they are used also on the threads of casing pipes. Other components treated include weapon parts, machine tool gears and, to achieve silent operation, the moving parts of refrigerator compressors.

3. Chromium oxide slurry coatings

In this process,[21] developed by Kaman Sciences Corp., a wear resistant layer, approximately 50 μm thick, containing silica and alumina particles bonded with chromium oxide is applied to components. The adherence and strength of the coating is high and a fine surface finish with good friction and wear properties can be produced.

(a) Coating process

A slurry, containing silica and alumina particles (<10 μm size) suspended in a concentrated chromic acid solution, is used. Components are coated and then heated to approximately 650°C. The porous ceramic layer is then impregnated with a saturated chromic acid solution and re-heat treated. This procedure is repeated up to twelve times and each heat treatment converts the chromic acid present to chromic oxide, Cr_2O_3. The resultant coatings are then claimed[22] to have hardnesses of 1800–2000 HV, ie close to the hardness of the alumina particles embedded in the coating.

A micrograph of a coating produced by Soudanor, Lille, France is shown in Fig. 7.5. The microhardness of this coating using a 50 g load was in the range 1700–1850 HV and using a 100 g load 450–650 HV. The low values using the 100 g load are due to the relatively porous structure fracturing under the diamond indenter. Using a 50 g load the indentation is smaller and can be placed in solid material between pores.[23]

101

Fig. 7.5 Micrograph of chromium oxide bonded coating produced by Soudanor. (X400)

(b) Properties and applications

The coating can be applied to alloy steels, iron, titanium and stainless steels but is preferably used on 13% chromium steel substrates. The bond strength is 80 MN/m² compressive strength 800 MN/m² and flexural strength 200–300 MN/m².

The coatings are reported to have excellent thermal shock resistance withstanding quenching into water from 600°C. The maximum working temperature is 700°C. Because of the inert nature of the constituents the ceramic layer can resist most corrosive fluids including sea-water, bases, most acids and solvents.

The process can be used to produce a durable outer layer on worn parts refurbished by, for example, plasma spraying. It has been used on pump parts such as linings, shafts, bodies, pistons, plungers and compressor rods and also on thrust washers, mechanical packings, sealing rings, moulds and extrusion or injection screws. Development work to use it on the linings of reciprocating engine cylinders and on parts for air bearings is presently being carried out.

A pump body in a catalytic cracking unit[22] hard-faced with Stellite 12 wore out after eight months. With the chromium oxide coating applied it has survived eighteen months without any wear. Other successful applications have been in a mechanical seal rubbing against a carbon bush and in the linings of chemical pumps and bearings working in corrosive and abrasive environments.

References

[1] PEARLSTEIN, F and WEIGHTMAN, R F. Electroless deposition of silver using dimethylamine borane, *Plating*, Vol 61, (2), p 154, 1974.

[2] *Electroplating of Plastics*, ed Weiner, R. Finishing Publications Ltd, Teddington, 1977.

[3] DUNCAN, R N. Engineering properties of sodium borohydride reduced electroless nickel deposits. *Electroless Nickel Conference III, Products Finishing*, Cincinnati, Ohio, 1983.

[4] GRAHAM, A H, LINDSAY, R W and READ, H J. The structure and mechanical properties of electroless nickel. *J Electrochem Soc.*, Vol 109, (12), p 1200, 1963 and Vol 112, (4), 1965.

[5] BEER, C F. Physical and engineering properties of electroless nickel deposits. *Recent Developments in the use of Electrodeposition Technology for Engineering Coatings*. Inst. Metal Finishing, Birmingham, 1983.

[6] *Electroless nickel, State of the Science*. National Association of Corrosion Engineers Task Group T-6A 53 Report, 9 October 1982.

[7] PARKER, K. Hardness and wear resistance tests of electroless nickel deposits. *Plating and Surface Finishing*, Vol 61, (9), p 834, 1974.

[8] TURNS, E W, BROWNING J W and JONES, R L. Electroplates on titanium; properties and effects, *Plating and Surface Finishing*, Vol 62, (65), p 443, 1975.

[9] HARRIS, S J, GOULD, A J and BODEN, P J. The effect of heat treatment and coating thickness on the fretting wear properties of electroless nickel in the temperature range 20–600°C. p 303. *Wear of Materials*, ed Ludema K C. ASME, New York, 1983.

[10] DINI, J W and JOHNSON, H R. Quantitative adhesion data for electroless nickel deposited on various substrates. *Electroless Nickel Conference III, Products Finishing*, Cincinnati, Ohio, 1983.

[11] GRAHAM, A H and GIBBS, T W. Wear of plated surfaces by synthetic fibres. *Symposium on properties of electrodeposits.* p 255. The Electrochem Soc, Princeton, NJ, 1975.

[12] FELDSTEIN, N, LANCSEK, T, LINDSAY, D and SALERNO, L. Electroless composite plating. *Metal Finishing*, p 35, August 1983.

[13] Information from the Montgomery Plating Company, Godiva Place, Coventry, CV1 5PN.

[14] Information from Ionic Plating Co Ltd, Grove Street, Smethwick, Warley, West Midlands B66 2QN.

[15] HUBBELL, F N. Chemically deposited composites—a new generation of electroless coatings. *Trans. Inst. Metal Finishing*, Vol 56, (2), p 65, 1978.

[16] LUKSCHANDEL, J. Diamond-containing electroless nickel coatings. *Trans. Inst. Metal Finishing,* Vol 56, (3), p 118, 1978.

[17] GUEGUEN, T. Improving the lubrication properties of ferrous metals. *Engineering Materials and Design*, June 1969.

[18] LORIN, G. *Phosphating of Metals*. Finishing Publications Ltd, Hampton Hill, Middlesex, 1974.

[19] RAUSCH, W. New phosphating lubricant for the cold forming of steel. *Metallgesellschaft AG Review, Surface Treatment Protection for Metals*, 19, Edition 19, 1977.

[20] KHALEGHI, M, GABE, D R and RICHARDSON, M O W. Characteristics of manganese phosphate coatings for wear-resistance applications. *Wear*, Vol 55, p 277, 1979.

[21] CHURCH, P K and KNUTSON, O J, (Kaman Sciences Corp), US Patent No 3,985,916, 1976.

[22] Soudanor technical literature. Soudanor, 87 Rue de la Plaine, 59000 Lille.

[23] MOORHOUSE, P, IRD. Private communication.

CHAPTER **8**

Chemical vapour deposition

1. Introduction

Chemical vapour deposition (CVD) is the thermally-induced decomposition or interaction of gaseous substances on a solid surface to produce a coating. Some diffusion into the substrate, and possibly reaction with constituents of the substrate, may also occur. The number of elements and compounds that can be deposited is large and many have found application in the field of solid-state electronics. Those used to improve wear resistance are titanium carbide (TiC), titanium nitride (TiN), titanium carbonitride Ti(C,N), alumina (Al_2O_3), tungsten carbide (W_2C and W_3C) and tungsten (W). Some information has been published on the deposition of chromium carbide (Cr_3C_2) silicon carbide (SiC)[1], hafnium nitride (HfN)[2] and titanium diboride (TiB_2)[3] but these materials have not as yet found commercial application; a multi-client programme to develop wear resistant coatings of cubic boron nitride (BN) was begun at Battelle Columbus Laboratories in 1981.

The largest application for wear resistant CVD coatings is on metal-cutting tools, about 50% of replaceable cemented carbide tool tips being coated in this way. Although the use of these coatings on high speed steel cutting tools is limited, they have been used on other steel tools, such as dies and cold extrusion punches. Metal coatings were the first to be produced by CVD, but engineering applications have arisen only for those that cannot easily be deposited by other methods, eg tungsten.

CVD processes are carried out at high temperatures, usually over 500°C and often around 1000°C. In general, sintered carbide substrates can tolerate such temperatures without distortion or loss of properties. Tool steels, however, require a subsequent hardening and tempering treatment and some distortion is almost inevitable. Because of the high temperature and consequent solid-state diffusion, CVD coatings are dense and highly adherent. Significant differences in thermal expansion coefficient cause interfacial stresses, and limit coating thickness usually to below 10 μm.

2. Deposition Procedure

(a) Equipment

The apparatus required for chemical vapour deposition comprises gas supply equipment, reaction vessel(s), heating system, vacuum pump and exhaust for waste gas.[4] A carrier gas, usually hydrogen, is passed through or over the liquid or solid reactant (generally a halide) to volatilise it, the temperature determining the concentration reached. If multiple layers are to be deposited, a number of gas streams is required. A plant with four reactors and two heaters used for the commercial coating of cemented carbide is shown in Fig. 8.1.

Articles to be coated are sometimes heated by induction, but most wear resistant coatings are deposited in externally heated hot-wall reactors. These are usually cylindrical in shape with components standing on trays supported by a central column and heated by radiation from the reactor wall. Uniform thermal and mass flow conditions are easier to attain in this than in a cold-wall reactor, so that coatings of uniform thickness can be applied to different materials and shapes in a single operation. Deposition may take place at atmospheric pressure, but low pressure operation has now become common practice as it improves quality and uniformity, and lowers deposition temperature. The reactants and their products are corrosive and the exhaust system therefore incorporates a scrubber and neutraliser. Corrosion must be borne in mind in selecting materials of construction to avoid plant failure and contamination of the coatings.

(b) Pre-treatment

Surfaces must be free from scale or other gross contamination. Preliminary exposure to a low pressure inert or reducing atmosphere at a temperature above 500°C removes minor contaminants. Problems with adhesion are more common if the deposition temperature is unusually low. Thorough cleaning is then advisable and it may be helpful to use an undercoat, nickel for example being used in bonding a coat of W_2C to steel.

(c) Post-treatment

With cemented carbide substrates, neither the carbide phases nor the cobalt matrix are significantly affected by the temperature used in CVD. Steels, on the other hand, require post-treatment to restore their optimum properties. Water and oil quenching are not practical immediately after the CVD process, but quenching with argon or nitrogen can be carried out within the reaction chamber. The usual procedure, however, is a separate austenitising, quenching and tempering cycle, carried out after the coating operation, preferably in a gas quench vacuum furnace. Steels that are successfully coated include the D2 and M2 types.

3. Deposition Reactions

The various types of CVD reaction are listed below, the most important for wear resistant coatings being (d), (e) and (f).

(a) Pyrolysis or thermal decomposition
$$Ni(CO)_4 = Ni + 4CO$$

(b) Reduction
$$WF_6 + 3H_2 = W + 6HF$$

(c) Oxidation
$$SiH_4 + O_2 = SiO_2 + 2H_2$$

(d) Hydrolysis
$$Al_2Cl_6 + 3CO_2 + 3H_2 = Al_2O_3 + 3CO + 6HCl$$

(e) Carbide formation
$$TiCl_4 + CH_4 = TiC + 4HCl$$

(f) Nitride formation
$$TiCl_4 + \tfrac{1}{2}N_2 + 2H_2 = TiN + 4HCl$$

(g) Co-reduction
$$TiCl_4 + 2BCl_3 + 5H_2 = TiB_2 + 10HCl$$

Of the above reactions those for producing nickel and tungsten occur at temperatures as low as 200 and 300°C respectively. However, at atmospheric pressure most of the reactions for producing compounds only proceed at useful rates at temperatures of approximately 1000°C. Lowering the pressure frequently increases reaction rate and allows the processes to be used at lower temperatures. A number of other methods of lowering the temperature of compound formation have been investigated, but none are used widely for the production of wear resistant coatings. For example CVD processes using metal organic compounds (MOCVD)[5] are widely used in the production at relatively low temperatures of compound semiconductors and other materials for the electronic industry, but are not used in the production of wear resistant coatings.

In producing titanium carbonitride coatings[6] the temperature can be lowered from 1000 to 650°C by substituting an amine for methane in the gas mixture. Other organic compounds containing carbon and nitrogen, eg nitriles, are also effective in lowering the deposition temperature.[7] Using ammonia instead of nitrogen to produce titanium nitride coatings allows a similar reduction in temperature. The coatings have a finer grain structure, and when applied on top of the ordinary high temperature deposit increase the resistance to abrasive wear.[8]

Another method of lowering the deposition temperature is to activate the reagents by an electrical discharge—plasma assisted chemical vapour deposition (PACVD).[9, 10] A dc or rf plasma is ignited around the components in a low-pressure atmosphere of argon, the temperature is raised to the required level and the reacting gases admitted. Deposition temperatures down to 550°C have been obtained for titanium carbide and as low as 300°C for titanium nitride and carbonitride, the lowest temperature attainable because of the heat generated by the plasma.[11]

4. Coatings

(i) Titanium carbide (TiC)

Titanium carbide is usually deposited by reaction between titanium tetrachloride and methane, with hydrogen as the carrier gas:

$$TiCl_4 + CH_4 = TiC + 4HCl$$

At atmospheric pressure and a temperature in the range 900–1200°C, a dense carbide coating is deposited at rates up to 5 μm/h. At pressures below 200 torr, these rates can be achieved at temperatures in the range 700–900°C; the thickness is also more uniform and the surface less rough (0.5 μm CLA).

Titanium carbide coatings are generally non-stoichiometric, with a significant excess of titanium. On sintered carbides or high-carbon steels, however, the composition may be close to stoichiometric, diffusion of carbon from the substrate making up any deficiency.

When the substrate is a cemented carbide, the depleted layer resulting from this diffusion contains η-carbide (Co_6W_6C) which may be several microns thick. This layer lowers the transverse rupture strength and significantly affects the toughness of cutting tools; its formation, however, can be largely prevented by a pre-carburising treatment.[12] The rate of carbon diffusion in the substrate exerts an effect on the rate of growth of CVD titanium carbide coatings. For example, deposits grow more rapidly on a cemented carbide with a high cobalt content than on one containing less cobalt; nucleation begins at the interface between the carbide particles and the matrix but then spreads much more rapidly over the cobalt, because of the higher carbon diffusion rate.[13] Similarly, on steels the rate of growth increases with carbon content and with nickel and cobalt content as these lower the activation energy for carbon diffusion. It decreases, however, in the presence of strong carbide formers (Cr, Mo, Ti) which raise the activation energy.[14]

The structure of CVD titanium carbide coatings is relatively coarse and columnar, facilitating crack propagation and reducing the transverse rupture strength[12] of cemented carbide substrates. Iron acts as a catalyst increasing the number of nucleation sites and so refining the grain.[15]

(ii) Titanium nitride (TiN)

The reaction generally employed for depositing titanium nitride is:

$$TiCl_4 + \tfrac{1}{2}N_2 + 2H_2 = TiN + 4HCl$$

hydrogen being used as carrier gas. In commercial practice deposition temperatures are in the range 550–950°C at pressures between 10 and 70 torr. Titanium nitride has a coarser columnar structure than titanium carbide but substituting ammonia for some of the nitrogen in the gaseous reactants and depositing at the lower end of the temperature range gives finer grained coatings.[10]

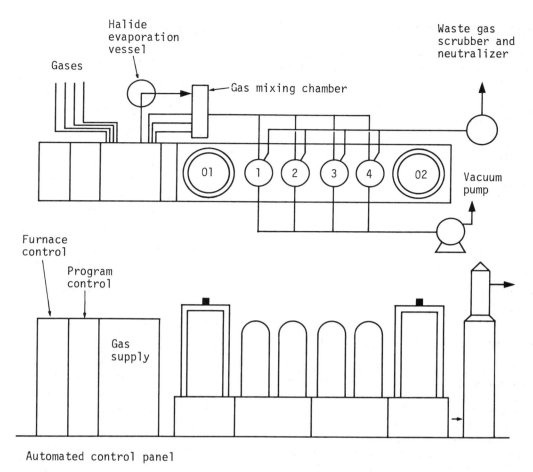

Fig. 8.1 Sketch of production CVD unit (1–4 coating vessels, 01 and 02 transportable furnaces).[4]

(iii) Titanium carbonitrides TiC_xN_{1-x}

The crystal structures of TiC and TiN are similar, and a complete range of mixed carbonitrides can be obtained. The composition of a coating can be changed through its thickness, either smoothly or discontinuously, by varying the proportions of the carbon- and nitrogen- containing constituents of the process gas. Deposition temperatures are in the range 700–900°C.[11]

(iv) Alumina (Al_2O_3)

Alumina is deposited by the reaction:

$$Al_2Cl_6 + 3CO_2 + 3H_2 = Al_2O_3 = 3CO + 6HCl$$

in the temperature range 850–1100°C.[16] [17] At the reduced pressure of 50 torr, the deposition rate varies with increasing temperature between 1 and 20 μm/h.

(v) Tungsten carbide (WC, W_2C and W_3C)

Deposition of WC has been claimed to result from reaction between tungsten hexafluoride and an aromatic hydrocarbon at 400–1000°C.[18] Formation of WC has also been reported from a mixture of tungsten hexafluoride, methane and hydrogen at 900–1100°C following initial deposition of W_2C and/or tungsten itself.[19] In further investigations,[20] however only the lower carbides W_2C and W_3C have been identified.

At temperatures below 700°C, a mixture of tungsten hexafluoride, hydrogen and methane yields about 90% metallic tungsten and very little carbide. Use of an aromatic instead of an aliphatic hydrocarbon results in a deposit of W_2C, benzene giving the highest deposition rate, 40 to 170 μm/h over the range 500–700°C.[20] Surface smoothness is favoured by low temperatures and pressures. A gas composition in which tungsten, carbon and hydrogen are in the molar ratio 1.5/1/1.7 results in a W_2C coating. Lowering the W/C ratio does not result in the formation of WC and raising it to two (stoichiometric for W_2C) results in the deposition of W_3C. Neither W_2C nor W_3C adheres well to steel, but bonding is improved when a nickel undercoat is used.[21]

In spite of the advantage of low deposition temperature, tungsten carbide coatings have been slow to find industrial applications. A commercial service has, however, been available since January 1984 supplying tungsten carbide coatings 2–30μm thick on a 0.5–5μm undercoat of nickel; the coatings are mainly W_2C if deposited above 400°C and W_3C if deposited at a lower temperature.[21]

(vi) Tungsten

The reaction used to deposit tungsten is:

$$WF_6 + 3H_2 = W + 6HF$$

105

A cold-wall reactor is used, the articles to be coated being heated by induction or resistance heating to a temperature in the range 300–700°C. The coating has a coarse, columnar structure, which becomes finer if the tungsten hexafluoride is injected close to the substrate; addition of carbon monoxide to the reactants gives a harder, striated deposit, probably containing some carbide. Coatings adhere firmly on carefully prepared steel, electron probe scans showing a 9 μm thick interdiffusion layer.[22]

5. Properties of coatings

Some properties of the materials used for CVD wear resistant coatings are given in Table 8.1. Considerable differences in hardness are quoted in the literature for a given material probably reflecting both variations in composition and structure and the inaccuracies in measurement of very hard materials. Average values are quoted. Thermal conductivity is relevant to the performance of coated cutting tools, low conductivity materials acting as thermal barriers and decreasing the working temperature of the carbide. Some values are shown in Table 8.2.[23]

The friction and wear properties of CVD titanium carbide coatings have been determined under various conditions with a view to developing roller bearings which can operate in the environments found in space and in nuclear reactors.[24] Some of the results obtained in a test in which a steel (100 Cr6) ball is pressed against a rotating disc of a similar steel with and without a titanium carbide coating are shown in Fig. 8.2. In the absence of lubrication, the titanium carbide layer markedly reduces the coefficient of friction both in air and in high vacuum. The wear rate of the uncoated steel ball against an uncoated steel disc is about 2000 times that against a coated disc. The wear rate of the uncoated steel disc is about twenty times that of the coated disc. In high vacuum there is complete seizure of the uncoated couple.

Table 8.1 Some properties of CVD wear resistant materials

Material	Microhardness HV	Coefficient of thermal expansion $10^{-6}/°C$ (20–1000°C)	Density g/cm³
TiC	3,200	7.5	4.93
TiN	2,000	9.4	5.22
TiCN	2,500–3,000	8.4	5.0
Al₂O₃	2,100	8.5	3.97
WC	2,400	5.4	15.63
W₂C	1,900	6.0	17.15
SiC	2,500	5.5	3.22
TiB₂	3,300	6.1	4.50
VC	2,600	7.2	5.77
Si₃N₄	2,800	2.8	3.44
B₄C	3,000	5.5	2.52

Table 8.2 Thermal conductivity as a function of temperature[23]

Material	Thermal Conductivity, W/m/°C					
	100°C	300°C	500°C	1000°C	1200°C	1500°C
TiC	33	35	37	41	42	44
TiN	21	22	23	26	27	28
Al₂O₃	28	18	13	6	5.5	5.5

Coatings of W_2C have been subjected to a range of wear tests which illustrate the limitations and advantages of hard brittle coatings on softer substrates.[21] Coatings of W_2C on high speed steel balls in a high load 4 ball wear test, where the main wear mechanism is rolling contact fatigue, flaked badly because of deformation of the substrate and low bond strength. However, in a low load abrasion test using ruby (alumina) abrasive, W_2C had one of the lowest wear rates of the materials tested (Table 8.3).[25] In another test in which cylinders of steel with various coatings were tumbled with an abrasive the depth of

Fig. 8.2 Coefficient of friction of steel against steel and steel against TiC under various conditions.[24]

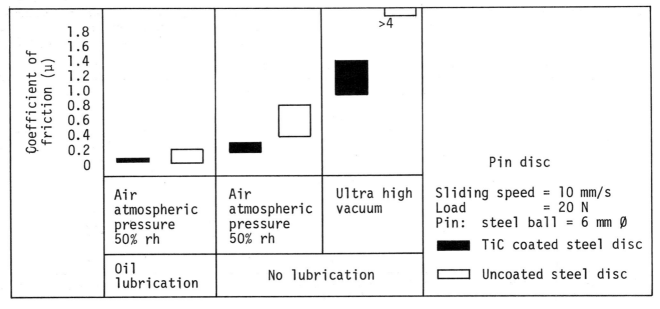

Table 8.3 Relative abrasive wear of various coated and uncoated material[25]

Material	Treatment or coating	Hardness HV	Relative Volumetric Wear
105WCr6	TiC (CVD at 1050°C), 12 μm	3,200	3.5
Cl00	W_2C (CVD at 550°C), 30 μm	1,900	4.8
100Cr6	Vanadised*, 30 μm	2,400	7.5
100Cr6	Borided†, 100 μm	1,600	14
CK15EH	Electrodeposited Cr, 50 μm	1050	23
—	Plasma sprayed NiCrBSi	700–800	33
X220CrMoV12H	Hardened, tempered at 150°C	810	48
—	Plasma sprayed Cr_3C_2/NiCr	500	50
AlMgSi	Hard anodised, 20 μm	450	63
100Cr6H	Hardened, tempered at 150°C	840	72

*Vanadised by pack cementation at 1100°C, 4h
†Borided by pack cementation at 900°C, 3h

wear was comparable with that of a D-gun coating (Table 8.4). Titanium carbonitride coatings produced at temperatures in the range 750–950°C, using acetonitrile or dimethylhydrazine, also have a low rate of wear in this type of test.[7]

6. Applications

The principal engineering application of CVD coatings is on cemented carbide cutting tools. Coatings extend tool life by increasing the resistance to crater wear on the rake face and predominantly abrasive wear on the flank face; they lower the cutting forces, making higher cutting speeds possible and reduce vibration, so improving the surface finish of the workpiece.[26]

Titanium carbide coatings, 2–3 μm thick, were the first to be used and remain the most important. The multi-layer coating, TiC/Ti(C,N)/TiN (1.5/1.5/2 μm) proved better in some applications, because TiN has greater thermodynamic stability and lower solubility in austenitic iron than TiC. Al_2O_3 is even more stable and very resistant to high machining temperatures, so coatings of TiC/Al_2O_3 followed. It is possibly advantageous also to place Al_2O_3 next to the substrate, if other layers are to follow, as it reduces carbon diffusion from the substrate to the coating. The relatively low thermal conductivity of alumina also reduces the temperature of the carbide substrate and

Table 8.4 Tumbling abrasive test results[21]

Material	Mean depth of wear (μm)	
	after 12 h	after 30 h
Sintered tungsten carbide	—	0.2
Electrodeposited Cr	—	0.5
D-gun tungsten carbide	—	1.2
CVD W_2C	0.2	2
CVD TiC	10	—
Nitrided NK7 steel	—	3.5

thus helps to maintain edge strength.[23] Proprietary combinations advertised include Sandvik's GC 415 (TiC/Al_2O_3/TiN) and Plansee's Sr 17 which consists of ten layers.[4] This application is discussed in Chapter 3.

All CVD hard coatings are brittle and liable to fracture on sharp edges. Tool tips designed for fine cuts are thus not suitable for coating. The roundness of the cutting tip on coated tools makes them unsuitable for machining nickel- or cobalt-based superalloys. CVD coatings are used to a limited extent on cemented carbide dies for wire and tube drawing.

The high deposition temperature of most CVD processes severely restricts their use on steel cutting tools. Even if the steel can be hardened and tempered after coating, some distortion inevitably occurs, and is likely to be unacceptable if tolerances are tight (less

Table 8.5 Some applications of CVD coatings on metal forming and cutting tools[27]

Tool type	Tool Material	Average life (operations)		Improvement
		uncoated	coated	
Draw die	D2	20,000	1,500,000	75.0X
Cut off tool	M2	150	1,000	6.3X
Trim die	M2	11,000	100,000	9.1X
Thread rolling die	D2	100,000	1,000,000	10.0X
Form tool	M2	4,950	23,000	4.6X
Extrusion punch	M2	2,000	70,000	35.0X
Tap	M2	3,000	9,000	3.0X
Class C hob	M2	1,500	4,500	3.0X
Drill	M2	1,000	4,000	4.0X
Minting collars	D2	200,000	1,000,000	5.0X

This table is reproduced from a leaflet issued by PFD Limited[28]. It includes some low precision cutting tools, which require a 3μm thick coating; the forming tools perform better with an 8 μm thick coating.

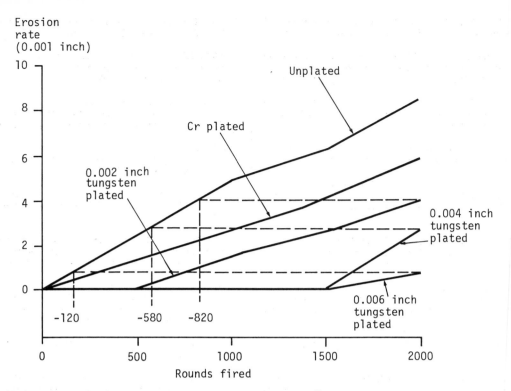

Erosion
rate
(0.001 inch)

Fig. 8.3 Erosion of unplated, chromium plated and tungsten plated gun barrels.[22]

than 0.02 mm) or the length to diameter ratio is greater than about fifteen to one. Where coated steel cutting tools are used, however, surface finish is improved and tool life extended.[26] On a range of cold-forming steel tools, CVD coatings of TiN, TiC or TiC/Ti(C,N)/TiN are very effective in increasing life. Some applications are shown in Table 8.5.[27]

Ball bearings coated with TiC have performed well in the focusing gear of the orbiting European Meteostat telescope, in nuclear reactors operating at 300°C in an atmosphere of dry helium, and in gyroscope motors lubricated with oil where coated races run more smoothly for longer periods.[1]

Titanium carbonitride coatings have been applied to compressor blades to protect them from erosion. Although no engine trials have been reported, the rate of erosion in laboratory tests was less than one-fifteenth that of any of the other materials tested, which included various steels, titanium and flame sprayed tungsten carbide.[6]

Metallic tungsten coatings have been used to resist erosion in the nozzles of solid propellant rockets and the bores of gun barrels.[27] Erosion rates for tungsten plated barrels—test fired at ten rounds per second, in fifty round bursts at one minute intervals—are compared with those for uncoated and chromium plated barrels in Fig. 8.3.

CVD tungsten coatings were highly resistant to wear when applied, with nickel undercoats, to ball valve seats in a coal gasifier; erosion tests showed that the coatings were twice as resistant as Stellite 6B at 20°C and five times as resistant at 550°C.[28]

References

[1] HINTERMANN, H E. Tribological and protective coatings by chemical vapour deposition. *Thin Solid Films*, Vol 84, p 215, 1981.

[2] KODAMA, M and BUNSHAH, R F. Interrupted cutting tests of cemented carbide tools coated by physical vapour deposition and chemical vapour deposition techniques. *Thin Solid Films*, 96, p 53, 1982.

[3] PIERSON, H O and MULLENDORE, A W. Thick boride coatings by chemical vapour deposition. *Thin Solid Films*, 95, p 99, 1982.

[4] SCHINTLMEISTER, W, WALLGRAM, W and KANZ, J. Properties, applications and manufacture of wear-resistant hard material coatings for tools. *Thin Solid Films*, 107, p 117, 1983.

[5] MOSS, R H. Metallo-organic compounds. *Chemistry in Britain*, 19, p 733, 1983.

[6] YAWS, C L and WAKEFIELD, G F. Scale-up process for erosion-resistant Ti(C,N) coating. *Proc. 4th Inter. CVD Conf., Electrochem Soc.* p 577, (1973).

[7] BONETTI-LANG, M, BONNETTI, R, HINTERMAN, H E and LOHMANN, D. Carbonitride coatings at moderate temperatures obtained from organic C/N compounds. *Proc. 8th Inter. CVD Conf., Electrochem Soc.* p 606, 1981.

[8] SJÖSTRAND, M E. Deposition of wear resistant TiN on cemented carbides using mixtures of NH_3/N_2 and $TiCl_4/H_2$. *Proc. 7th Inter. CVD Conf. Electrochem Soc.*, p 452, 1975.

[9] ARCHER, N J. Plasma assisted chemical vapour deposition. *Surfacing Journal*, Vol 14 (1), p 8, 1983.

[10] HOLLAHAN, J R and ROSLER, R S. p 335, IV–1, *Thin Film Processes*, ed J L Vossen and W Kern, Academic Press, New York, 1978.

[11] ARCHER, N J. The plasma-assisted chemical vapour deposition of TiC, TiN and TiC_xN_{1-x}, *Thin Solid Films*, 80, p 221, 1981.

[12] SARIN, V K and LINDSTROM, J N. The effect of eta phase on the properties of CVD TiC-coated cemented carbide cutting tools. *Proc. 6th CVD Conf., Electrochem Soc.*, p 389, 1977.

[13] LEE, C W and CHUN, J S. Nucleation and growth of CVD TiC coatings on cemented carbides. *Proc. 8th Inter. CVD Conf., Electrochem Soc.*, p 540, p 544, 1981.

[14] ROSER, K. TiC growth studies on different standard steels. *Proc. 8th Inter. CVD Conf., Electrochem Soc.*, p 586, 1981.

[15] HINTERMAN, H E, GASS, H and LINDSTROM, J N. Nucleation and catalysed growth of TiC on cemented carbide. *Proc. 3rd Inter. CVD Conf., The American Nuclear Society*, p 353, 1972.

[16] LINDSTROM, J N and JOHANNESSON, R T. Nucleation of Al_2O_3 layers on cemented carbide tools. *Proc. 5th Inter. CVD Conf., Electrochem Soc.*, p 453, 1975.

[17] FUNK, R, SCHACHNER, H, TRIQUET, C, KORNMANN, M and LUX, B. Coating of cemented carbide cutting tools with alumina by chemical vapour deposition. *Proc. 5th Inter. CVD Conf., Electrochem Soc.*, p 469, 1975.

[18] AUTOMOTIVE PRODUCTS CO. LIMITED., British Patent No. 1,326,797, 1973.

[19] MANTLE, H, GASS, H and HINTERMANN, H E. Chemical vapour deposition of tungsten carbide (WC). *5th Inter. CVD Conf., Electrochem Soc.*, p 540, 1975.

[20] ARCHER, N J. Tungsten carbide coatings on steel. *5th Inter. CVD Conf., Electrochem Soc.*, p 556, 1975.

[21] ARCHER, N J and YEE, K K. Chemical vapour deposited tungsten carbide wear-resistant coatings formed at low temperatures. *Wear*, 48, p 237, 1978.

[22] MAYER, K H. CVD tungsten as a means to reduce high temperature erosion in gun barrel bores. *Proc. 3rd Inter. CVD Conf., The American Nuclear Society*, p 625, 1972.

[23] PORAT, R. Thermal properties of coating materials and their effect on the efficiency of coated cutting tools. *Proc. 8th Inter. CVD Conf., Electrochem Soc.*, p 533, 1981.

[24] HINTERMANN, H E. Exploitation of wear and corrosion-resistant CVD coatings. *Tribology International*, Vol 13, (6), p 267, 1980.

[25] STAHLI, G and BEUTLER, H. Evaluation wearing performance under abrasive and adhesive sliding by means of model tests. *Sulzer Tech Rev.* Vol 58, (1), p 33, 1976.

[26] SCHINTLMEISTER, W, KANZ, J and WALLGRAM, W. Possibilities and limitations for the application of CVD-coated cemented carbide tools and steel tools. *Inter. J. of Refractory and Hard Metals*, Vol 2 (1), 1983.

[27] Trade Literature from PFD Limited. 27 Roman Way, Coleshill, Birmingham B46 1HQ

[28] STEPHENSON, J B and McDONALD, H O. Gasification pilot plant evaluation of CVD coated hardware. *Proc. 8th Inter. CVD Conf., Electrochem Soc.*, p 642, 1981.

CHAPTER 9

Physical vapour deposition

1. Introduction

In physical vapour deposition (PVD) processes material is volatilised in a low pressure chamber by heating (evaporation) or by bombardment with ions (sputtering). Pressures vary between 10^{-6} and 10^{-1} torr depending on the particular process. The temperature of deposition is usually below 500°C and coating adherence depends on cleanliness of the substrate and the kinetic energy of the material being deposited. PVD processes can be classified (Table 9.1) according to the type of source and whether there is:

(a) chemical reaction in the gas phase;
(b) ion bombardment of the article being coated;
(c) enhancement of the chemical reactivity of the gas phase by an electrical discharge.

Thermal evaporation and condensation is the simplest PVD coating process. It is used in optical, electronic and decorative applications. Coating adhesion is usually low but can sometimes be increased by post coating heat treatment.

Sputtering produces more adherent coatings which are mainly used in the same applications as evaporated coatings but also on some engineering components. It is fundamentally a slower process, but the development of magnetron and rf sputtering techniques has increased production rates.

Reactive evaporation and sputtering techniques in which the evaporated metal combines with a gas have been devised specifically for the deposition of compound coatings, some of which are hard and wear resistant, while others are effective solid lubricants.

Ion plating, in which a glow discharge is generated around the workpiece, gives a uniform adherent coating on complex shaped components. It was initially developed for the deposition of coatings to protect high strength steels against corrosion without risk of embrittlement. Reactive ion plating has more recently come into prominence as a method of applying wear resistant coatings, especially titanium nitride, to high speed steel cutting tools; reactive bias sputtering (or sputter ion plating) is also used for the same purpose.

Ion beam processes have not been used to any significant extent to deposit wear resistant coatings. More energetic ion beams are used in the process of 'ion implantation' which produces a dispersion of implanted atoms in a shallow surface layer rather than a coating.

All PVD processes are costly in terms of equipment and energy requirements are high. They present no environmental problems, however, and are capable of precise control.

Table 9.1 Physical vapour deposition processes

Process designation	Source			Chemical reaction	Activation of gas	Ion bombardment of substrate	Substrate penetration
	Thermal evaporation	Atom ejection (sputtering)	Ion beam generation				
Vacuum coating	✓						
Reactive evaporation	✓			✓			
Activated reactive evaporation	✓			✓	✓		
Sputter coating		✓					
Reactive sputtering		✓		✓			
Reactive bias sputtering		✓		✓		✓	
Ion plating	✓					✓	
Reactive ion plating	✓			✓		✓	
Biased activated reactive evaporation	✓			✓	✓	✓	
Primary ion beam deposition			✓			✓	
Secondary ion beam deposition	✓		✓				
Ion implantation			✓			✓	✓

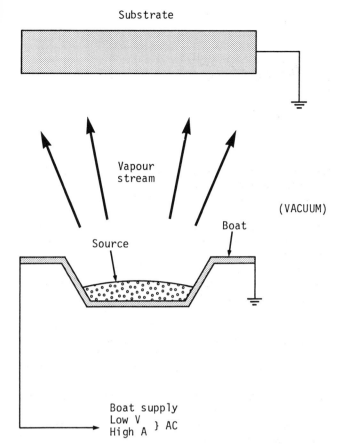

Fig. 9.1   Resistance-heated evaporation source.[4]

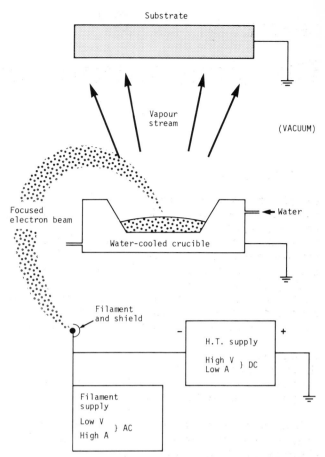

Fig. 9.2   Electron-beam heated vacuum evaporation source.[4]

2. Evaporation processes

(a) Operating principles

The rate of evaporation of a material is a function of its vapour pressure and the ambient pressure. When the total pressure is below 10^{-5} torr, metals evaporate at practical rates if their vapour pressures exceed 10^{-2} torr, ie at the temperatures shown in Table 9.2.[1] Such low ambient pressures avoid contamination of the coatings, but the mean free path

Table 9.2   Temperatures at which the vapour pressure of some metals reaches 10^{-2} torr

Metal	Temperature, °C
Ag	1030
Al	1220
Au	1400
Cd	265
Co	1520
Cr	1400
Cu	1260
Fe	1480
Hf	2400
Ni	1530
Mo	2530
Pb	715
Ti	1740
W	3230
Zn	345

of a particle is several metres, so the evaporant moves in a straight line between source and substrate. Clearly, good distribution is difficult to achieve on other than flat substrates, and complex movement of shaped workpieces is necessary.

Direct evaporation of ceramic materials is difficult because of their low vapour pressure and tendency to dissociate. Reactive methods are therefore employed to obtain coatings of carbides, nitrides or oxides, the metal being evaporated in the presence of an appropriate gas, such as methane, nitrogen or oxygen. Reaction to form the desired compound occurs on the workpiece and other surfaces. In activated reactive evaporation (ARE), the gas, at a pressure in the range $10^{-4} - 10^{-3}$ torr, is made more reactive by exciting it with an electric discharge.[2, 3]

(b) Process technology

The temperature required for evaporation is most simply achieved by resistance heating (Fig. 9.1). This is suitable for materials that vaporise below 1600°C and are inert to the container, which may be made of tungsten, molybdenum or tantalum. Induction heating is feasible, if a suitable crucible can be found, but is not a favoured technique, partly because of the risk of arcing. Electron beam heating, in spite of its cost and complexity, is now the most popular method. Equipment used is shown diagrammatically in Fig. 9.2.[4] By holding the evaporant in a water-cooled copper crucible compatibility problems are

111

avoided. The electron gun may be of the hot filament type, as shown, or of the hot or cold hollow cathode type. The beam of electrons is usually curved through 270° and focussed to a spot which is scanned across the source producing a pool of molten evaporant. The guns are designed to operate in high vacuum so that a two-zone chamber may be required. The aperture through which the beam passes offers high resistance to the flow of gas, and it is possible to maintain a pressure ratio of 100 to 1 between the two zones. Heating of the substrate, if required, can be by radiant heaters or by intermittent deflection of the electron beam. Arc sources have characteristics making them particularly suitable for ion plating systems, and are described later.

Alloys may sometimes be evaporated from a single source but, if the vapour pressures of the constituents differ substantially, multiple sources are necessary to control deposit composition.

Fig 9.3 shows the configuration adopted for activa-

ted reactive evaporation. Ionisation within the plasma increases the chemical reactivity of the gas, but the substrate lies outside the plasma and is not subject to ion bombardment.

(c) Components

To obtain good bonding, articles to be coated should first be thoroughly cleaned and degreased. Preheating is desirable to increase the rate of outgassing of the substrate and is essential for thick metal and ceramic coatings to minimise interfacial stresses. As vapour deposition in a high vacuum is a line-of-sight process, it is difficult to obtain uniform coating thickness on irregular surfaces.

(d) Coatings

Most elements can be evaporated; some metals, eg silver, lead and cobalt, are used in tribological applications. Metallic coatings more than a few microns thick, deposited at low temperature, are slightly

Fig. 9.3 The ARE process with an electron beam evaporation source.[3]

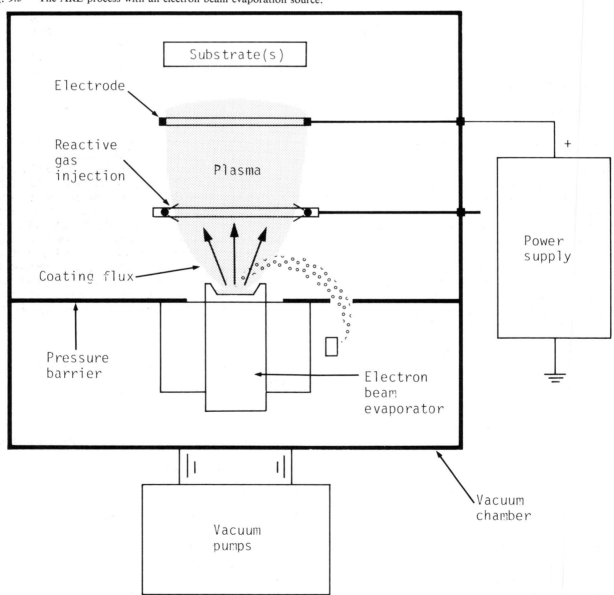

porous, have a columnar structure with low intercrystalline strength, and are poorly bonded to the substrate. If the temperature can be raised to above half the melting point of the coating(°K), diffusion and grain growth give rise to stronger adhesion and an equiaxed structure.

Coating rates in vacuum evaporation[5] can be as high as 75 μm/min. Such rates are useful in special circumstances, eg coating strip or producing foil, but rates of 2 or 3 μm/min are more normal for applying coatings to shaped articles.

Many refractory compounds have been deposited by reactive evaporation; some are listed in Table 9.3. Titanium carbide formed by reacting titanium vapour with acetylene at 4×10^{-4} torr can be deposited at rates between 0.6 and 10 μm/min depending on the rate of evaporation of the metal. The effects of substrate temperature on the rate of deposition and deposit hardness are given in Table 9.4. Although grain growth occurs at the higher temperatures, the deposits remain columnar.[6]

(e) Applications
Applications of evaporated coatings are listed in Table 9.5. Wear resistance is a factor in only a few cases.

3. Sputtering processes

(a) Operating principles
When a glow discharge is formed between two electrodes in a gas pressure in the range $10^{-3} - 10^{-1}$ torr and at an applied potential above about 500 V, positive ions bombard the negative electrode (cathode or target) and remove material by momentum transfer. The material removed, mainly in the form of neutral target atoms, condenses on solid surfaces in its path to give a coating. Virtually any material can be sputtered. If it is a compound, decomposition is likely to occur and precautions have to be taken to achieve a coating of the same composition. With alloys, differential sputtering of the various components may occur initially until a steady-state condition is reached, in which the target surface has a lower proportion of the constituents with the higher

Table 9.3 Coating rates of refractory compounds by reactive and activated reactive evaporation[2]

Coating process	Coating rate, μm/min	Compound type
Reactive	<0.2	Oxides: Al, Si, Ti Nitrides: Al, Nb, Ta, W
Activated reactive Outside*	<0.2	Oxides: Al, Si, Ti Nitrides: Al
Inside*	1.5 – 10	Nitrides: Ti Carbides: Hf, Nb, Ta, Ti, V, Zr

*Refers to gas activated inside or outside the reaction zone

Table 9.4 Effect of substrate temperature on rate of deposition and hardness of TiC on tantalum[6]

Substrate temperature, °C	Thickness of deposit, μm	Rate of deposition, μm/min	Deposit hardness, $HV_{0.05}$
520	100.0	3.3	2710
610	93.0	3.1	2330
730	87.5	2.9	2955
830	67.5	2.6	2955
1080	37.5	2.2	4110
1120	30.0	2.15	4160
1450	55.0	1.85	3890

sputtering yield; subsequently, the coating composition will be close to that of the target.

Energies of the bombarding ions are in the range 100–1000 eV, and of the sputtering atoms 10–40 eV. Such energetic particles form stronger bonds when they strike the substrate than evaporated atoms, which typically have energies of only 0.2–0.3 eV. The bombarding ions are generally those of a heavy inert gas, argon being the most common. Up to 0.1 torr, the sputtering rate rises with gas pressure, because the ion concentration increases, but at higher pressures it falls as a result of back-scattering. The sputtering rate increases with current density but only slightly with voltage.

Table 9.5 Applications of vacuum evaporated coatings

Application	Purpose
Electronic components	Manufacturing procedure
Interference films on camera and microscope lenses	To reduce reflection and glare
Coloured interference coatings	For decoration, absorption of solar energy, control of heating, eg of satellites
Coatings on architectural glass	Tinting, to reduce heat and glare
Transparent coatings, eg alumina, silicon dioxide, on polycarbonate windows and eye-glasses	To give abrasion resistance
MCrAlY coatings on gas turbine vanes and blades (M = Co, Ni or Fe)	For increased resistance to sulphidation and oxidation
Aluminium coating of steel strip	For use in can manufacture
Nickel coating of perforated steel strip	For use in alkaline battery manufacture
Coating of interior and exterior car trim	For decoration

(b) Process technology

The basic sputtering system is a simple diode (Fig. 9.4). The target is connected to a negative voltage supply and faces the substrate at a distance of about 10 cm or less. When a glow discharge forms in low pressure argon at a potential in the range 0.5–5 kV, positive ions bombard the target and produce a stream of vaporised particles, many of which condense on the substrate. The coating rate in these circumstances is between 0.5 and 5 μm/h. Use of a dc voltage limits the process to conductors, but recourse to rf excitation extends its application to semiconducting and insulating materials, and increases ionisation and hence the deposition rate.

To raise the rate of deposition substantially it is necessary to increase the ionisation further. Triode systems incorporating thermionic electrodes have been used, but have largely been superseded by magnetrons in which magnetic fields confine the plasma and enhance the ionisation efficiency by increasing the path length of the electrons and the number of collisions they undergo. The most important configurations are planar and cylindrical. Planar magnetrons are convenient but inherently non-uniform sources, whereas cylindrical magnetrons with a central target give uniform vapour streams. Hollow-cathode magnetrons are also used for coating complex shapes. In general, permanent magnets are preferred to electromagnets. Although developed initially for dc systems, magnetrons can be adapted for use with an rf supply. Using magnetrons metal deposition rates up to 100 μm/h are possible.

Ceramic coatings can sometimes be deposited by direct sputtering, but a reactive gas addition may be required to maintain the composition of compounds that suffer dissociation. Alternatively, a reactive gas may be used in conjunction with a sputtered metal to give the required compound at the surface of the substrate. However, reaction also occurs on the target and if a continuous coating forms, the sputtering rate drops sharply, leading to loss of control. In the reactive sputtering of ceramic materials deposition rates are usually below 10 μm/h. Some recent work however, has shown that under laboratory conditions, the reactive sputtering rate can be raised close to that of the metal if the process is precisely controlled. If

Fig. 9.4   Sputter coating process.[5]

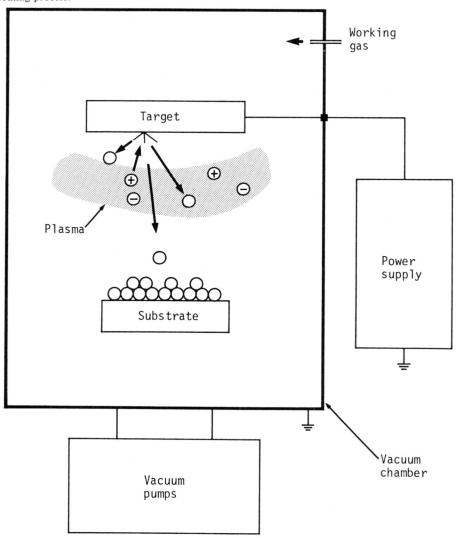

translated into industrial practice this advance could significantly improve the economics of the process.[7]

If a negative potential is applied to the workpiece, the process is known as bias sputtering. Deposition is then accompanied by ion bombardment of the workpiece, which increases the density and purity of metal coatings and refines the structure of both metal and ceramic deposits. Adhesion also increases, especially as the bombardment is usually started before deposition begins and provides a period of sputter cleaning. The bias potential is usually in the range 100–500 V.

A simple form of lightly-biased sputtering equipment, usable in the reactive mode, was developed at AERE, Harwell[8] and is now manufactured by Tube Investments plc.[9] Its main features are shown in Fig. 9.5. It is a dc unit with no magnetrons but has a target which almost covers the inside vertical walls of a heated enclosure within the vacuum chamber. A simple mechanical pump is adequate to maintain the vacuum required ($10^{-2} - 10^{-1}$ torr). Heating the enclosure to about 300°C and purifying the incoming gases keeps the impurity concentration as low as in high vacuum systems. The heater is switched off when sputter coating begins as the glow discharge is able to maintain the working temperature. The deposition rate is low, up to a few μm/h, but adherent, dense coatings are obtained. Because the target surrounds the work and the pressure is relatively high, throwing power is good and complex articles can be plated without mechanical manipulation.

An inverse configuration is employed in the hot rod rf sputtering system devised by Dowty Electronics plc.[10] The target is in the form of a rod with work-pieces arranged around it, and during operation it attains a temperature of approximately 1000°C. In reactive sputtering of zirconium and titanium nitrides, the compounds form on the target surface, but at this temperature they can be sputtered directly at reasonably high and constant rates. It is claimed that uniform coating compositions can thus be achieved without elaborate gas flow controls; workpieces are not biased, there are no magnetrons and overall power requirements are low at 0.4 kW per target. The coating rate is low, however, at about 1 μm/h. Two 13-target, fast cycle machines, commissioned in 1984, are suitable for treating a wide variety of components.

A plant of a different type, designed for on-line production coating of cutting tools with titanium nitride, uses high substrate biasing and magnetrons, and operates at a deposition temperature above 400°C to achieve the adhesion and structure required. A double-diode arrangement with flat targets above and below the components ensures uniformity of thickness and composition, without the need for rotation. Using input and output vacuum interlocks, with preheating in the input chamber, it is possible to deposit 2–5 μm TiN coatings in a cycle time of 40–60 min.[11]

(c) Components

Coatings can be deposited on metallic and non-metallic materials, including ceramics, plastics, paper and fabrics. Pre-cleaning is necessary, vapour degreasing being usual for metal components. Sputter cleaning is accomplished in dc diodes and highly biased systems by making the workpiece the cathode in

Fig. 9.5 Bias sputtering (or 'sputter ion plating') system.[8]

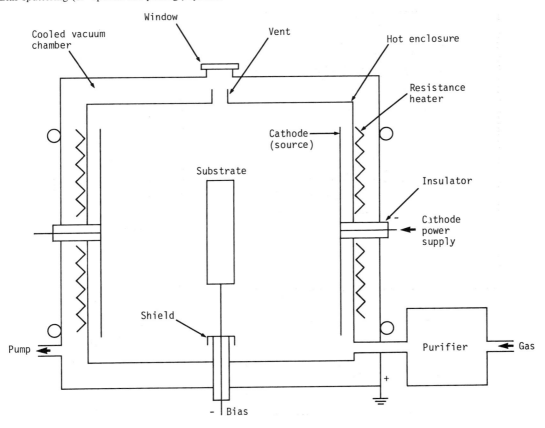

an electric discharge. During this initial period it is advisable to interpose a screen to prevent the sputtered material from contaminating the target. If sputter cleaning is not possible a preheat stage may be used to remove residual films. An elevated deposition temperature is desirable to improve structure and bonding.

(d) Coatings

Sputtering is one of the most versatile of coating methods in that virtually any material that does not decompose under ion bombardment can be deposited. Many compounds that decompose can nevertheless be deposited by reactive sputtering. The process is therefore widely used to produce coatings of metals, alloys and compounds such as oxides, carbides, nitrides and some sulphides.

The structure of 25 μm thick metallic coatings, deposited at rates of 6–12 μm/h, consists of columnar grains and is similar to that of evaporated films. The formation of equiaxed grains requires a substrate temperature as high as 80% of the melting point.[12] Gases, especially argon, may be trapped in sputtered films, but biasing the substrate and raising the temperature tend to reduce the amount.

The microhardness of reactively sputtered 4–5 μm thick coatings of TiN, on the surface of stainless steel, varied between 2200 and 3000 HV depending on the nitrogen partial pressure.[11] Similar values were obtained on thicker coatings deposited at a high rate, (Table 9.6) and are considerably higher than values for bulk material.[7] A similar effect was found with ZrN.

A study of the friction coefficient and wear resistance of sputtered and ion-plated TiN deposits on cast iron substrates has been carried out.[13] The heterogeneous structure of the cast iron produced defects in the coatings, and creates some uncertainty about the results. Table 9.7 and Fig. 9.6, however, compare the most consistent values obtained with those for chromium plated and uncoated specimens. The tests were performed on pin specimens rubbing on a lubricated cast iron disc. Friction values are similar to the low values obtained with cast iron-cast iron and cast iron-electroplated chromium couples whilst the wear rate of the TiN coated pins is much less.

Dry lubricant films can be applied by sputtering. An investigation of directly sputtered MoS_2 coatings has shown that fibrous coatings break at the base of the fibres during sliding but the film remaining, which is only about 0.2 μm thick, provides effective lubrication.[14] Co-sputtering up to 10% of a metal such as nickel with MoS_2 is reported to lower and stabilise the coefficient of friction.[15]

(e) Applications

Sputtering competes with evaporation in many of the applications shown in Table 9.5, especially in the manufacture of electronic components, and the coating of architectural glass. Razor blades have been coated with Cr_3Pt to improve hardness and corrosion resistance. Sputtered TiN coatings are applied on a large scale to watch cases to produce a simulated gold finish of high durability.

Table 9.6 Microhardness of TiN and ZrN deposited by very high rate reactive sputtering[7]

Coating	Thickness μm	Load, gf	HV Optical	HV SEM	Bulk hardness, HV
TiN	9.5	100	3238	3359	2000
	9.5	200	2663	2391	2000
ZrN	11.0	100	3342	2628	1500
	11.0	200	2358	2663	1500

Table 9.7 Friction coefficient of TiN coatings on cast iron substrates[13]

Time, h	0	2	5	10	20
Uncoated	0.09	0.10	0.10	0.08	0.08
Electrodeposited Cr	0.11	0.12	0.11	0.09	0.09
Sputtered TiN	0.15	0.12	0.10	0.09	0.08
Ion plated TiN	0.11	0.10	0.10	0.10	0.10

The sputtering of chromium onto plastic car radiator grilles was discontinued after several years because of inadequate durability; the plant continued in operation, however, applying a metallic film as a base for subsequent electroplating.

The use of both sputtered and ion plated TiN on steel cutting tools began only a few years ago but is becoming of increasing importance. This application has been discussed in Chapter 3.

4. Ion plating

(a) Operating principles

Ion plating is usually performed in argon at a pressure of about 10^{-2} torr. A negative potential in the range 2–5 kV is applied to the workpiece so that it becomes the cathode of a glow discharge formed between it and the earthed parts of the apparatus. The consequent argon ion bombardment, which first cleans the surface, is continued while coating material is evaporated into the plasma and deposited on the substrate. This continuing bombardment increases nucleation and improves adhesion by enhanced diffusion and impingement mixing. Gas scattering is the major factor in ensuring uniform coating distribution but the shape of the electric field and some re-sputtering may also help. It was formerly thought that many of the evaporated atoms were ionised and impelled towards the substrate with maximum energy, but it was later realised that the ionisation efficiency is generally less than 1%.[16] With arc evaporators however, higher levels of ionisation can be obtained.[17]

In reactive ion plating, the argon is partly or completely replaced by a gas capable of reacting with the evaporant to give the desired coating material.

(b) Process technology

The equipment used in the basic dc ion plating process is sketched in Fig. 9.7;[18] resistance heating, however,

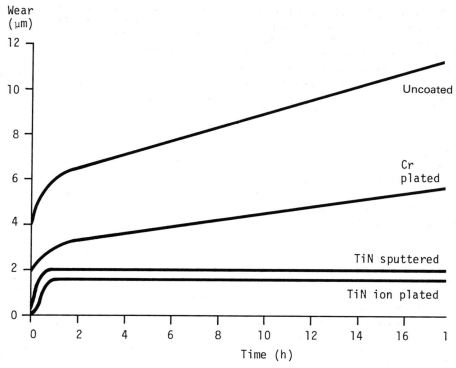

Fig. 9.6   Wear behaviour of TiN coatings on cast iron substrate.[13]

Fig. 9.7   Ion plating apparatus.[18]

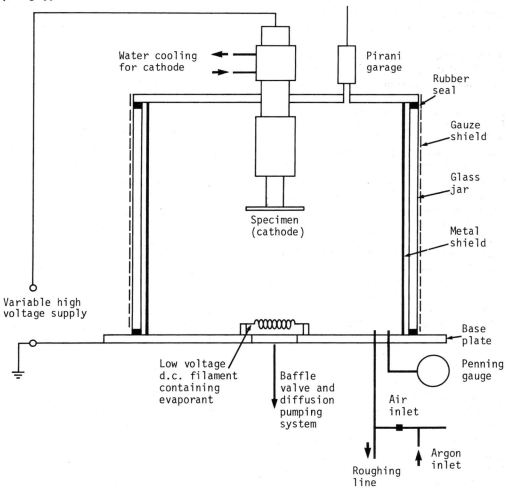

is often replaced by an electron beam. An rf supply is sometimes employed and is essential if the substrate is non-conducting. Reactive gases may be introduced to form compound coatings. If a subsidiary discharge is employed to increase reactivity, the process is essentially the same as biased activated reactive evaporation (BARE).

An important development in ion plating has been the introduction of multi-arc evaporation sources, especially for the reactive plating of high speed steel tools. Each arc is in continuous movement, so that rapid evaporation occurs without formation of a pool of liquid. Thus, the sources can face in any direction, as illustrated in Fig. 9.8,[19] unlike a molten source, which has to face upwards. It is therefore easier to obtain good coverage on a full load of shaped components. The plasma used in this version of the ion plating process, to clean the substrate and continue bombardment during reactive deposition of TiN, is unusual in comprising mainly titanium ions and electrons. Metal coatings can be deposited at high rates by ion plating; 300 μm/h being readily achieved in laboratory equipment. Reactive deposition of coatings like TiN, are carried out at much lower rates ($\sim$10 μm/h).

(c) Components

It is advisable to pre-clean substrates. Vapour degreasing is required on metal components and is sometimes followed by ultrasonic alkaline cleaning, rinsing and drying. Sputter cleaning, which precedes ion plating itself, removes residual surface films.

The temperature continues to rise during deposition, limiting the coating thickness that can be applied to certain plastics, eg 2.5 μm of stainless steel was chosen as a safe maximum on acrylonitrile butadiene styrene (ABS). In other circumstances, the temperature rise may be helpful, eg in producing a diffusion zone between titanium and an aluminium deposit.[20]

(d) Coatings

Ion plating has been used to produce a wide range of metallic and non-metallic coatings. Compared with those deposited without ion bombardment of the workpiece, they have superior adhesion and a less columnar structure when produced at low temperatures. Surfaces are also smoother. In practice coating thicknesses are limited by the rate of coating formation. Metal coatings over 100 μm thick have been reported but most compound coatings are in the 1–10 μm range. Thicker compound coatings can, however, be produced using longer deposition times.

Soft metals (eg Cu, Ag, Au, Pb) for use as dry lubricants have been deposited by ion plating.[21] Some results of pin on disc tests carried out in vacuum on gold and lead coated stainless steel (440C) are shown in Fig 9.9. In each case there is a minimum coefficient of friction at a coating thickness in the range 0.1–0.2 μm.

Ion plated cobalt has poor tribological properties because its crystal structure is predominantly face centred cubic. Cobalt-chromium alloy coatings, on the other hand, are entirely hexagonal close packed.

Fig. 9.8   Multiple arc ion plating system.[19]

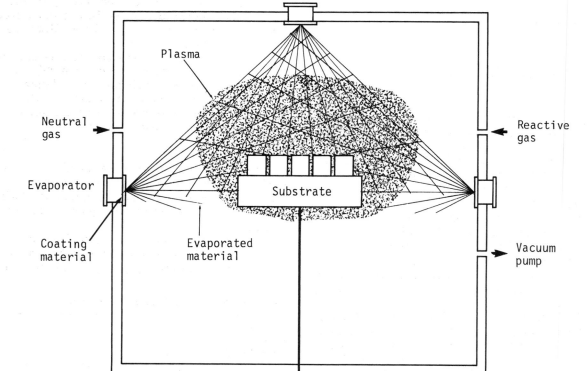

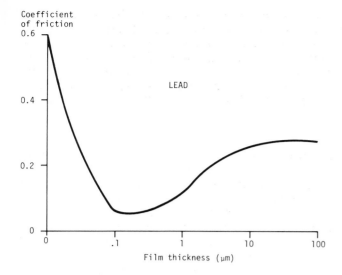

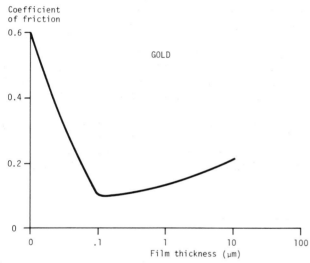

Fig. 9.9   Variation in friction coefficient with thickness of, (a) lead and (b) gold, ion plated coatings.[21]

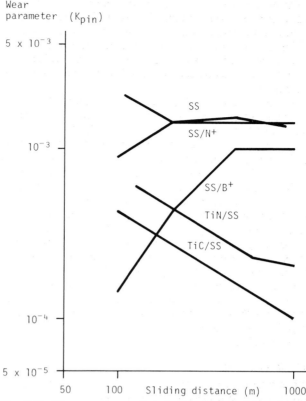

Fig. 9.10   Wear parameter of pin as function of sliding distance in dry tests.[20]

Their properties depend, however, on the intensity of the ion current. For example, the microhardness has been observed to be 408 ± 36 HV at 0.15 mA/cm^2, but 1421 ± 228 HV at 0.3 mA/cm^2. The friction coefficient and rate of wear in pin on disc tests were also much higher in the former case.[22]

Ion plated aluminium bronze on steel forming tools has been investigated.[23] Again the properties depend on the ion current density as shown in Table 9.8. In production scale trials a coating which had only been pregreased extended the life of a collaring pin by a factor of three compared with an uncoated pin which had been both pregreased and lubricated continuously during use.

Fig. 9.10 summarises the results of pin on disc tests without lubrication, where the pin is 440C stainless steel and the disc, which is also 440C, is either untreated, nitrogen or boron ion implanted, TiC or TiN coated (using biased activated reactive evaporation). The benefit of ion implantation soon disappears, whereas that of the TiC and TiN coatings persists. TiC and TiN coatings also exhibit superior resistance to abrasion by 600 grit silicon carbide powder with kerosene on nylon cloth. The behaviour in erosion tests depends on the severity of the conditions: they protect at low levels, but suffer accelerated brittle failure at the higher levels. Both coatings exhibited strong adhesion.[20]

Diamond-like carbon films can be produced from hydrocarbon atmospheres by various highly energetic deposition processes, including ion plating. These films have a high hydrogen content, a density of 1.8 g/cm^3, a high resistivity and are transparent. Their hardness[24] is approximately 600 HV.

(e) Applications

Ion plated aluminium (or occasionally cadmium) is used on steel and titanium fasteners to provide corrosion resistance in aerospace applications. MCrAlY deposits on superalloys protect against high temperature corrosion.

Ion plating can provide intermediate coatings to help in joining (eg silver on aluminium) or in applying adherent electrodeposits (eg copper on titanium).

Table 9.8   Properties of ion plated aluminium bronze[23]

Coating number	1	2	3	4
Ion current density, mA/cm^2	0	0.1	1	2.5
Microhardness, HV	180	230	305	165
Sliding distance to wear through 5 μm, m	0*	60	>250	90

*Failed during running-in period.

119

As mentioned above, ion plated soft metals are used as dry lubricants, eg on bearing surfaces for use in space and vacuum equipment.

The application of hard coatings to high speed steel cutting tools, which is one of the important growth areas in ion plating, has been discussed in Chapter 3.

5. Ion beam processes

(i) Primary and secondary ion beam deposition

In primary ion beam deposition, the beam comprises relatively low energy(~ 100 eV) ions of the desired film material which deposits directly onto the substrate. Complete and independent control can be exercised over film composition, particle energy and area coated (since the beam can be deflected and focussed). In secondary ion beam deposition, more energetic beams (~ 1 keV) are directed onto a target, which is sputtered onto a substrate. Compared with an ordinary sputtering process, this procedure offers operation at lower pressure together with independent control of direction, energy and target material. Compounds such as oxides and nitrides can also be sputtered reactively.[25]

In a process called 'dynamic recoil mixing', one ion beam sputters a film onto a substrate, while another bombards the film to produce mixing at the interface, with consequent benefit to the chemical, mechanical or electrical properties.[26]

(ii) Ion implantation

(a) Process technology

Ion implantation is carried out in vacuum ($10^{-6} - 10^{-5}$ torr) by directing a beam of ions, accelerated to an energy around 100 keV, at the workpiece. The ions penetrate to an average depth of about 0.1 μm and a maximum depth of 0.25 μm. The technique is widely used to inject dopants into the surface of semiconductors, there being more than 1500 implanters in use for this purpose throughout the world. The treatment of engineering components requires much higher doses (about 100 times) and special configurations to cope with more complex geometry. Based on a prototype developed by AERE, Harwell, suitable equipment is now commercially available,[27] with a chamber about 0.6 m cube and a beam current of 5 mA. It is estimated that with this it should be possible to treat 100 cm^2 of surface at a cost of £50/h (50p/cm^2, assuming 1 h cycle time), but that with scale-up this cost might be brought down to 10p/cm^2 over three years.[28] AERE, Harwell, have already brought into operation a facility with a chamber 2.5 m in diameter and 2.5 m long and a beam current of 10–20 mA of nitrogen ions.

Ion implantation may be used on coated as well as uncoated materials. The effect may simply be to modify the properties of the coating but it can also be used specifically to force atoms from the coating into the substrate by ion bombardment, as an alternative to their direct implanation. This process is known as ion beam mixing.

(b) Properties of treated surfaces

Ion beam deposition produces the same range of coatings as reactive bias sputtering or ion plating, with added advantages of versatility, accuracy and control. The extra complexity and cost have not greatly impeded its use in the electronics field, but have so far prevented the development of significant applications in engineering.

Ion implantation has, however, been used in some engineering applications. It has several advantages over coating techniques: pre-cleaning is not important, temperatures can be kept low, there is no change in dimensions and very little change in surface finish. It is, however, strictly a line-of-sight process.

Because of the shallow penetration of the ions their hardening effect cannot be measured by conventional

Table 9.9 Applications of ion implantation on tools[30]

Tool	Material	Implanted species	Factor of improvement
Injection moulding tool for plastics	Cr-plated steel	Nitrogen	10
Injection screw for plastics	Nitrided steel	Nitrogen	10–15
Feed wear pads and sprue bushes for plastics machinery	Steel	Nitrogen	18
Punch/die for motor laminations	Steel and cemented carbide	Nitrogen	6
Drills for foil-coated laminate	Cemented carbide	Nitrogen	2–3
Tools for metal extrusion	Steel, stellite, cemented carbide	Nitrogen	3
Ring press tool for steel	Steel	Nitrogen	10
Forming tool for steel	Carburised mild steel	Nitrogen	2–3
Hot rolls for non-ferrous rod	H13 steel	Nitrogen	5
Dies for copper rod	Cemented carbide	Carbon	5
Ring cutter for tinplate	Steel	Nitrogen	3
Gear cutter	Steel	Nitrogen	3
Thread cutting dies	M2 steel	Nitrogen	5
Swaging dies for steel	Cemented carbide	Nitrogen	2
Wire drawing dies	Cemented carbide	Carbon	3
Slitters for synthetic rubber	Cemented carbide	Nitrogen	12
Slitters for paper	1.6% Cr, 1% C steel	Nitrogen	2

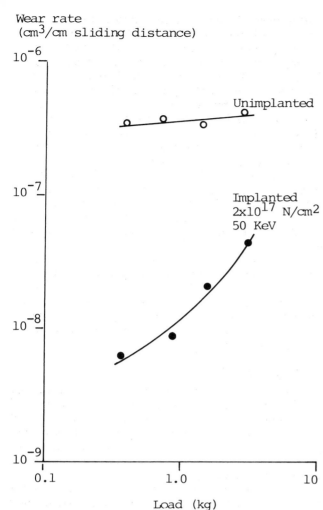

Fig. 9.11   Wear between a stainless steel pin and implanted or untreated En 40B steel disc as a function of the applied load.[30]

Fig. 9.12   The factor by which the wear rate is reduced as a function of nitrogen dose.[30]

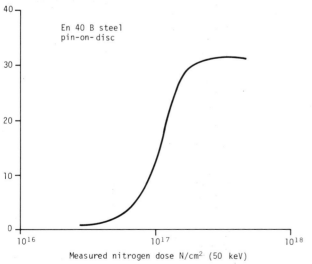

indentation techniques. Their effect on the wear resistance of En 40B in a pin on disc test is shown in Fig. 9.11, and Fig. 9.12 shows how the wear resistance in such tests varies with the nitrogen dose.[29]

During ion implantation compressive stresses are introduced which have a favourable effect on fatigue strength. For example, nitrogen implantations of 2×10^{17} ions/cm^2 into steel or titanium increase the number of cycles to failure in Wohler tests by 8 to 12 times.

(c) Applications

Although the commercial exploitation of ion implantation is limited, its ability to improve wear resistance under industrial conditions has been demonstrated for a wide range of items. Table 9.9 summarises the advantages resulting from the treatment of various tools. In another application, ion implantation of an extrusion tool for PVC enabled the material to be changed from a tool steel to a low carbon steel.

Other applications which are being explored include the implantation of chromium into ball bearings and nitrogen into stainless steel for pump and valve components.[30]

References

[1] RUDZKI, G J. Surface finishing systems. Amer Soc for Metals, p 149, 1983.

[2] BUNSHAH, R F and RAGHURAM, A C. Activated reactive evaporation process for high rate deposition of compounds. J. Vac. Sci. Tech. Vol 9, (6), p 1385, 1972.

[3] BUNSHAH, R F. Processes of the activated reactive evaporation type and their tribological applications. Thin Solid Films, Vol 107, p 21, 1983.

[4] BUCKLOW, I A. Physical vapour deposition techniques. Surfacing J. Vol 8, (4), p 3, 1977.

[5] BUNSHAH, R F. Deposition technologies for films and coatings—developments and applications (by R F Bunshah et al). Noyes Data Corp, Park Ridge, NJ, 1982.

[6] BUNSHAH, R F and RAGHURAM, A C. The effect of substrate temperature on the structure of titanium carbide deposited by activated reactive evaporation. J. Vac. Sci. Tech, 9., Vol 9, (6), p 1389, 1972.

[7] SPROUL, W D. Very high rate reactive sputtering of TiN, ZrN and HfN. Thin Solid Films, Vol 107, p 141, 1983.

[8] DUGDALE, R A. Sputter ion plating. Proc. Int. Conference on Ion Plating & Allied Techniques, IPAT 77, Edinburgh, June 1977.

[9] Information from Tube Investments plc, Hinxton Hall, Hinxton, Saffron Walden, Essex, CB10 1RH.

[10] DUCKWORTH, R G. High purity sputtered tribological coatings,. Thin Solid Films, Vol 86, p 213, 1981.

[11] MÜNZ, W D, HOFMANN, D, and HARTIG, K. A high rate sputtering process for the formation of hard friction reducing TiN coatings on tools. Thin Solid Films, Vol 96, p 79, 1982.

[12] THORNTON, J A. Influence of apparatus geometry and deposition conditions on the structure and topography of thick sputtered coatings. J. Vac. Sci. Tech., Vol 11, (4), p 666, 1974.

[13] KLOOS, K H, BROSZEIT, E, GABRIEL, H M, and SHRODER, H J. Thin TiN coatings deposited onto nodular cast iron by ion- and plasma-assisted coating techniques. *Thin Solid Films*, Vol 96, p 67, 1982.

[14] SPALVINS, T. Morphological and frictional behaviour of sputtered MoS_2. *Thin Solid Films*, Vol 96, p 17, 1982.

[15] STUPP, B C. Synergistic effects of metals ion sputtered with MoS_2. *Thin Solid Films*, Vol 84, p 257, 1981.

[16] TEER, D G. The ion plating technique. *Surfacing J.*, Vol 8, (4), p 11, 1977.

[17] BRANDOLF, H. Carbide and nitride wear coatings. *11th Int. Conference on Metallurgical Coatings*, American Vacuum Society, San Diego, Abstract 13. 1984.

[18] TEER, D G. Ion plating. *Trans. Inst. Met. Finishing*, Vol 54, (4), p 159, 1976.

[19] Information from Multi-Arc (Europe) Limited, Number One Industrial Estate, Medomsley Road, Consett, Co Durham, DH8 6SX

[20] SURI, A K, NIMMAGADDA, R, and BUNSHAH R F. Influence of ion implantation and overlay coatings on various physico-mechanical and wear properties of stainless steel, titanium and aluminium. *Thin Solid Films*, Vol 64, p 191, 1979.

[21] SPALVINS, T, and BUZEK, B. Frictional and morphological characteristics of ion plated soft metallic films. *Thin Solid Films*, Vol 84, p 267, 1981.

[22] ARNELL, R D, and TEER D G. Tribological properties of highly oriented Co and Co–Cr ion platings. *Thin Solid Films*, Vol 84, p 281, 1981.

[23] SUNDQUIST, H A, and MYLLYLA, J. Wear of ion plated aluminium bronze coatings. *Thin Solid Films*, Vol 84, p 289, 1981.

[24] ANGUS, J C. *et al*. Composition and properties of the so-called 'Diamond-like' amorphous carbon films. *11th Int. Conference on Metallurgical Coatings*, American Vacuum Society, San Diego, 1984.

[25] HARPER, J M E. *Thin Film Processes*, ed Vossen, J L and Kern, E. Academic Press, New York, p 175, 1978.

[26] Information from Department of Electronic and Electrical Engineering, University of Salford, Salford, M5 4WT.

[27] Manufacturers: Tecvac Limited, Main Street, Stow-cum-Quy, Cambridge, CB5 9AB

[28] DEARNALEY, G. The modification of materials by ion implantation. *Phys. Technol.*, Vol 14, p 225, 1983.

[29] DEARNALEY, G. The ion implantation of metals and engineering materials. *Trans Inst Met Finishing*, Vol 56, (1), p 25, 1978.

[30] DEARNALEY, G. Practical applications of ion implantation. *J. Metals*, Vol 34, (9), p 18, 1982.

Additional reading

Proceedings of 5th International Conference on Ion and Plasma Assisted Techniques. Munich. (Publisher: CEP Consultants Ltd 26 Albany Street, Edinburgh EH1 3QH), May 1985.

MATTHEWS, A. Titanium nitride PVD coating technology. *Surface Engineering*, Vol 1, (2), 1985.

CHAPTER 10
Sprayed coatings

1. Introduction

A number of processes have been developed in which fine particles of the coating material are injected into a gas stream, which is heated either electrically or by combustion, and deposited at high velocities on relatively cold substrates.

The main processes, Table 10.1, are:

powder or thermo spraying;
wire spraying;
detonation gun (or D-gun) spraying; and
plasma spraying.

The term flame spraying is frequently used loosely to describe all spraying processes.

Spraying processes have the following features:

(a) a wide variety of materials can be applied.

(b) coatings can be applied to a wide range of substrate materials. (Temperatures during spraying can usually be maintained below 200°C).

(c) there is little restriction on the shape of the component apart from satisfying the condition that the angle of impingement of sprayed material should be close to 90° and certainly not less than 60°C. Special guns are available for spraying bores over about 5 cm diameter.

(d) the size of components is only restricted by the handling equipment available.

Some of the limitations of sprayed coatings are:

(a) the relatively low bond strength. Flame and plasma coatings are considerably less well bonded to substrates than chromium plate or fused coatings. The bond strength of some D-gun coatings is approaching that of chromium plate to steel. In general sprayed metals and cermets such as tungsten carbide-cobalt are better bonded than sprayed oxides.

(b) the porosity in the coatings. Coating porosity can vary from about 20% in flame sprayed oxide coatings to about 1% in D-gun and some plasma coatings. This can allow corrosion of the substrate in aggressive environments.

(c) the thickness of the coating. Coating thicknesses are limited by stresses at the interface arising from thermal contraction of the coating. These stresses are least in porous ductile metallic coatings and greatest in dense, brittle ceramic coatings where coating thicknesses can be limited to 0.25 mm. The geometry of the part also affects permissible coating thickness. Greater thicknesses can be tolerated on cylindrical than on flat surfaces.

The structure of sprayed coatings is lamellar, consisting of particles which have been flattened by impingement at high velocity and high temperature on the substrate. When oxidation of the particles occurs during spraying a high concentration of oxide occurs at the lamellae boundaries. However in processes in which oxidation is prevented little trace of the original lamellar structure is visible on microscopic examination. Sprayed metals are usually harder than the corresponding wrought metal, mainly because of the presence of dispersed oxide formed during spraying. They are also less ductile because of the increased hardness and the porosity.

2. Processes

(i) Surface pretreatments

In all spraying processes it is essential that the substrate be suitably prepared to ensure good bonding with the coating. This preparation consists of machining, cleaning, surface roughening and possibly application of a bond coat.

(a) Machining

It is necessary, especially with thick coatings as deposited by wire spraying to undercut, ie machine undersize, the area being coated to allow for the required coating thickness.

(b) Cleaning

The part must be cleaned of dirt, oil and grease. This is usually done in a vapour degreasing unit using trichlorethylene.

(c) Surface roughening

This can be done by cutting a rough V thread or by using a rounded grooving tool. In both cases the grooves are approximately 0.5 mm deep. Grooving is preferred for thick (>0.75 mm) metallic coatings applied by wire spraying if a bond coat is not used.

However the most commonly used method for most spraying processes is grit blasting. The grit is usually an impure alumina and both pressure blasting and suction blasting can be used. With thin plasma coatings the surface roughness of the as-sprayed coating is affected by the grit blasted surface. It is therefore necessary to apply only a light grit blast where minimum as-sprayed surface roughness is required. For highest bond strength it is necessary to spray within two hours of grit blasting. It is difficult to roughen a surface sufficiently to obtain a good bond when the substrate hardness is greater than 55 HRC(600HV).

(d) Application of bond coat

Certain materials, notably molybdenum and mixtures of nickel and aluminium, bond much more readily to most substrate materials than other coating materials. The nickel-aluminium mixture is available as powder for flame and plasma spraying and as wire for wire spraying. As it is less operator dependent than molybdenum and also gives a higher bond strength it is now widely used. On clean as-machined surfaces bond strengths of approximately 28 N/mm^2 (4000 lb/in^2) can be obtained using nickel-aluminium coatings but they are more frequently used on grooved or grit blasted surfaces to improve the bonding of thick coatings applied by wire spraying; the surface of as-sprayed nickel-aluminium coatings, being very rough, provides a good bonding surface for subsequent coatings. In general the bond strength of plasma sprayed coatings is adequate without the use of bond coats but they are sometimes used to improve bonding to hard substrates, or as a build-up coat before application of a material which is thickness limited.

(e) Preheating

It is also desirable to preheat the surface to around 100°C to ensure that it is dry before spraying. It is claimed that in some systems there is an optimum preheat temperature for maximum bond strength. (see section 3).

(ii) Combustion processes

(a) Powder or thermo-spraying

In thermospraying a simple oxyacetylene gas torch, into which the spray powder is aspirated, is used. The main advantages of the process are low capital cost and ease of operation. It is mainly used for restoring the dimensions of components which have inadvertently been machined undersize. Because of the small size of the thermospray gun and the ease with which it can be manipulated there is little restriction on the size or complexity of components which can be sprayed.

However particle velocities are low and consequently coating porosities are high (~20%). Bond coats of nickel aluminide (coating 2 Table 10.2) are used. A number of powder mixtures have been developed which when applied by thermospraying give a low porosity coating well bonded to the substrate. They are referred to as self bonding, one step coatings, as no separate bond coat is required. Their common feature is a relatively high aluminium content (5-15%) which on spraying reacts exothermically with other constituents in the powder. Two examples of such powders (1 and 3) are given in Table 10.2. However these powders are expensive and for this reason are only used for restoration work where the demand is small and intermittent. Otherwise the wire process is used. Ceramic materials such as alumina, chromium oxide and zirconia are also thermosprayed, an initial bond coat of nickel aluminide or nickel chromium alloy being used. Coating porosities are high (~20%) and bond strengths relatively low. Coatings are used in low load sliding wear applications and as thermal barriers.

(b) Thermo-spraying with subsequent fusion

Coatings which are subsequently fused in a separate operation are also deposited by the thermospray gun. The most widely used have been developed from a nickel chromium solid solution containing 10-20% Cr, the melting point of which has been lowered to 1000-1100°C by the addition of small amounts of carbon, boron and silicon. These elements also contribute to the hardness by forming hard chro-

Table 10.1 Comparison of principal spraying processes

Heating Method	Combustion			Electric Arc		
Process	Powder	Wire	D-gun	Wire	Plasma (40 KW)	Plasma (80 KW)
Approx. gas temp. °K	3,000	3,000	4,000	5,000	15,000	15,000
Approx. particle velocity, m/s*	35	200	800	200	300	500
Typical bond strengths, N/mm^2 (lb/in^2)	45 (3,000)	45 (3,000)	145-360 (10,000-25,000)	45 (3,000)	45-120 (3,000-8,000)	60-145 (4,000-10,000)
Typical coating porosities, %	2	10	1	10	2	1
Materials sprayed	Metals, ceramics	Metal wires, ceramic rods	Metal, cermet and ceramic powders	Metal wires	Metal, cermet and ceramic powders	
Rates of deposition, kg/h	2-9	6 (metal) 1 (ceramic)	~1	15	1-5	1-5
Approx. cost of basic spraying equipment, £	3,600	4,500	Not sold	10,000	Superseded by higher power equipment	75,000
Typical coating thickness, mm	0.75	1.25	0.25	1.25	0.25	0.25

*Varies with particle size and density.

mium carbides, borides and silicides. In general the hardness of the alloy increases and the melting point decreases with increasing C + B + Si content. The boron and silicon also combine with oxygen on the substrate surface and on the surface of the sprayed particles during fusion and thus aid coalescence of the coating and bonding to the substrate. The high nickel and chromium contents ensure good oxidation resistance of the fused coatings up to about 950°C. Although cobalt-base alloys have been developed their use is confined to applications where hot hardnesses or improved corrosion resistance is required. Some examples of spray and fuse coatings are given in Table 10.3.

The Ni–Cr–B–Si alloys have excellent resistance to low stress abrasive wear but less to high stress grinding abrasion. They also have excellent resistance to sliding metal to metal wear. Being fully dense they have excellent corrosion resistance and find their greatest application in situations demanding both wear and

Table 10.2 Thermosprayed coatings

Coating	Nickel base 'self-bonding' alloy	Nickel aluminium	Aluminium bronze
Composition, %	Al 5.5 Mo 5 Ni bal	Al 5 Ni 95	Cu 85 Al 15
Hardness, HV	140	120	95
Bond strength to grit blasted steel, N/mm² (psi)	35 (5000)	20 (3000)	14 (2000)
Density, g/ml	7.2	7.2	6.5
Porosity %	<2	2	<5
Maximum service temp. °C	800	800	230
Thickness, mm typical	2.5	None	2.5
Cost, £/kg	42	22	33
Special features	Expensive material but used on account of high bond strength for reclamation work on difficult geometries. Can be machined to fine finish. Moderate wear resistance.	Used mainly as a bond coat. Can be used as a single coating to give moderate wear resistance up to 800°C. Poor corrosion resistance.	Mixture of elemental powders which react on spraying to give good bond. Used mainly as build up coating on copper base alloys. Good machining characteristics. Moderate wear resistance.

Table 10.3 Thermosprayed coatings (sprayed and fused)

Coating	Nickel base coating (hard)	Nickel base coating (soft)	Cobalt base coating	Nickel base alloy with WC particles
Composition, %	Cr 13.5 C 0.75 B 3.00 Si 4.25 Ni bal	Cr 10.0 C 0.45 B 2.00 Si 2.25 Ni bal	Ni 28 Cr 18 B 3.0 Si. 3.5 Co bal	(WC/8% Ni) 35 Cr 11.00 C. 0.5 B 2.5 Ni bal
Hardness, HV	700	400	520	950 (WC particles 2000)
Density, g/ml	7.8	8.2	8.5	10.2
Porosity, %	0	0	0	0
Melting point, °C	1040	1110	1120	1025

Thickness, mm
  upper limit
  lower limit
	1.75
	0.25
		1.75
		0.25
			1.75
			0.25
				None
				0.25

Cost, £/kg	13.5	13.5	27	51
Abrasive wear resistance relative to plasma sprayed WC/12% Co = 1	0.35	0.20	0.30	0.75
Special features	Standard hard facing material. Must be ground. Excellent wear and corrosion resistance.	Moderate wear resistance. Good impact properties. Can be machined with WC tools.	Good sliding wear resistance. Low coefficient of friction. Retains hardness at elevated temperature.	Outstanding wear resistance. Coating can be ground and lapped to fine finish. Expensive material.

Fig. 10.1 Principal spraying processes

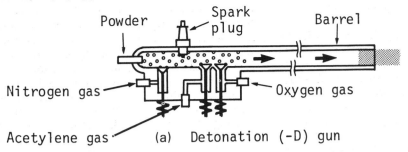

(a) Detonation (-D) gun

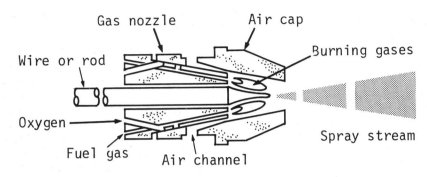

(b) Combustion wire gun

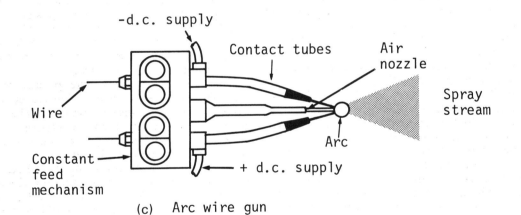

(c) Arc wire gun

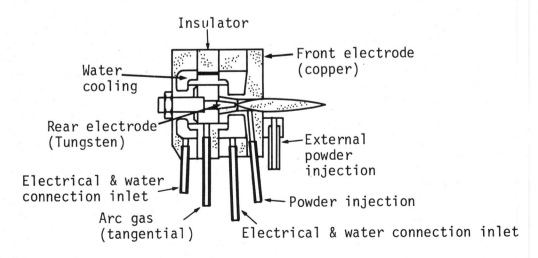

(d) Plasma-gun

corrosion resistance, eg in pumps, and in some high temperature applications, eg plungers in glass bottle making.

A further development of spray fuse alloys is the incorporation of comparatively large (100–150 μm) particles of tungsten carbide-cobalt or more recently tungsten carbide-nickel in a base Ni–Cr–B–Si alloy. The percentage of tungsten carbide-nickel particles can vary in the range 35–50%, the composition of a widely used material containing 35% tungsten carbide-nickel particles being shown in Table 10.3. These materials are expensive but they have extremely high resistance to abrasive particles, eg in slush pumps and exhaust fans, and also to heavy metal to metal wear, eg in wire-drawing capstans.

Fusion of the coatings is usually carried out using oxyacetylene, oxy-propane or oxy-hydrogen torches. R.F. heating may also be used. With components of variable section there is the danger, when fusing coatings on thicker areas, of overheating thinner areas with consequent 'running' of the coatings. Large components of uniform section can however be fused satisfactorily using torches. For example Ni–Cr–B–Si coatings approximately 1.5 mm thick are fused on continuous casting billet run-out rolls 3 m long, 0.5 m diameter and 60 mm wall thickness using large oxyacetylene burners.

Controlled atmosphere furnaces are also used for large components. Precise temperature control is necessary to prevent the coating running from the surface but provided this is achieved better quality control can be exercised than with torch fusion.

During fusion, shrinkage of approximately 20% occurs and the coating is metallurgically bonded to the substrate. Because of the low ductility and comparatively high expansion coefficient ($\sim 14 \times 10^{-6}/^{\circ}$C) of the fused coatings it is important that they be cooled slowly from the fusing temperature to avoid cracking. Controlled cooling is also necessary to obviate undesirable metallurgical changes in the substrate.

The spray-fuse process gives zero porosity of the coating and very high bond strengths to the substrate. The advantages over other welding processes are the high rate of deposition and the smoothness and uniformity of the deposit which reduces the amount of subsequent grinding required. The disadvantage of the spray-fuse technique compared with other spraying processes is that the substrate is heated to comparatively high temperatures (1050–1100°C) so that some distortion may take place and reclamation is difficult if a part is damaged during processing. In many applications spray-fuse coatings have been displaced by more wear resistant plasma sprayed tungsten carbide-cobalt coatings. These are less costly because of the smaller amount of material used, the reduced amount of machining needed and the elimination of the subsequent fusion process.

(c) Fuse welding

A process similar to spray-fusing is known as fuse-welding. This process is carried out with a small oxyacetylene torch into which powder is fed through a finger-operated valve. Powder is fed into the gas stream intermittently and fused in the intervals between feeding. The powders used are very similar to those used in spray-fusing, being nickel- or cobalt-based containing boron and silicon and melting around 1050°C. Some also contain fine tungsten carbide particles. Whereas the spray and fuse process is ideally suited to cylindrical components, the fuse welding process is more suited to irregular shapes, edges and thin sections such as are found in dies, bottle moulds and valve seats. The surface of the coating is usually rougher than in the conventional spray and fuse process and is more operator dependent. The rate of powder deposition in fuse welding is low (1 kg/h) and the process is normally used only on small irregular areas.

(d) Detonation or D-gun process[1]

The D-gun consists of a water cooled steel tube (approximately 2.5 cm internal bore and 1.4 m long) closed at one end (Fig.10.1a). At the closed end is a sparking plug and a system of valves through which are metered the appropriate quantities of acetylene, oxygen and the powder to be sprayed. The gas mixture is ignited either 4.3 or 8.6 times per second, depending on the type of coating being applied. The temperature of the gas issuing from the gun is about 3000°K and the particles have a velocity of about 800 m/s. The gun-to-substrate distance is in the range 5–10 cm. Molten or partially molten particles are deposited as a series of overlapping circular deposits approximately 50 μm diameter and 6 μm thick. The combustion gas can be neutral, oxidising or reducing and its temperature can be controlled by addition of nitrogen (to cool) or hydrogen (to heat).

The noise level during the process is very high (150 dB) so the equipment is housed in sound-proof double walled concrete cubicles. The process is remotely controlled from a console outside the cubicle, components being mechanically rotated and/or traversed in front of the gun.

The rate of deposition for a 250 μm coating is either 48 or 96 mm^2/s, depending on whether the gun is operated at 4.3 or 8.6 detonations per second. In the case of alumina this would correspond to a deposition rate of either $\frac{1}{4}$ or $\frac{1}{2}$ kg/h (cf. plasma spraying at approximately 2 kg/h).

Union Carbide Corporation is the sole operator of the process and has one or more plants in most Western industrialised countries. The prices charged for coating components are considerably higher (approximately by a factor of 2) than by plasma spraying and this can be accounted for by the comparatively high cost of the installation and low spray rate. The process is mainly used to coat components which are to operate under conditions of severe abrasive or percussive wear where the abrasion resistance or the bond strength of plasma sprayed coatings is inadequate.

The main D-gun coatings are either oxides or carbide-metal cermets, some of which are listed in Table 10.4. The coating structure, like that of other sprayed coatings, consists of a series of laminations parallel to the substrate surface. The porosity is usually low ($\frac{1}{2}$–1%) and the reported coating substrate bond strength for tungsten carbide-cobalt coatings is above 170 N/mm^2 (25,000 lb/in^2). This value is more than twice that reported for the corresponding plasma sprayed coatings.

(e) Wire spraying

A wire gun is illustrated in Fig.10.1(b). In this process, wire in the range 1.25–4.75 mm diameter is continuously fed into an oxyacetylene flame where it is melted and atomised by an auxiliary blast of compressed air.

A range of wire spray guns capable of varying deposition rates are available. The smaller guns have a wire feed mechanism, powered by a small compressed air driven turbine and can be hand held or machine mounted. In the larger guns, the wire feed is by an electric motor and these guns are almost invariably machine mounted. For spraying most materials an approximately neutral flame gives optimum values of deposition rate and bond strength. It is desirable that substrates be preheated to approximately 100°C before spraying to eliminate the possibility of condensation of water and also to reduce stresses between coating and substrate due to thermal mismatch. Preheating to minimise interfacial stresses is especially important in spraying internal bores.

Metal spraying is used primarily for building up worn or mismachined components. Most of the sprayed deposits have considerably better wear resistance than the original material, presumably because of the presence of oxide particles and the improved oil retention properties due to the presence of porosity.

Table 10.4 Typical D-gun coatings

Union Carbide coating designation	LA2	LW1	LW IN40	LC1B
Composition, %	Al$_2$O$_3$ 99	WC 91 Co 9	WC 85 Co 15	Cr$_3$C$_2$ 65 Ni-Cr 35
Hardness, HV	1100	1300	1050	700
Bond strength, N/mm^2 (lb/in^2)	70 (10,000)	140 (~20,000)	140 (~20,000)	140 (~20,000)
Density, g/ml	3.4	14.2	13.2	6.5
Porosity, % (by metallography)	2	$\frac{1}{2}$	$\frac{1}{2}$	1
Surface finish, μm CLA				
as-coated	4	4	3	4
ground and lapped	0.05	0.025	0.025	0.05
Special features	Good wear resistance but inferior to carbides. Chemically inert. Electrical insulator.	Highest wear resistance of all coatings. Limited to temperatures below 500°C. Moderate impact resistance.	Wear resistance slightly less than LW1 but higher impact and fretting resistance. Used on mid span stiffeners in jet engine compressor blades. Limited to below 500°C.	Can be used up to 850°C. Mainly used in temperature range 500–850°C where it has excellent wear resistance. Used in seals in turbine section of jet engines.

Table 10.5 Wire sprayed coatings

Coating	Molybdenum	Nickel-aluminium	Stainless steel	Aluminium bronze
Composition, %	Mo 99.9	Ni 80 Al 20	Cr 13.4 C 0.32 Fe bal	Cu 90 Al 9 Fe 1
Hardness, HV	400	250	350	155
Bond strength, N/mm^2 (lb/in^2)	17 (2500)	28 (4000)	Bond coat usually used	
Density, g/ml	8.86	6.00	6.74	7.06
Porosity, %	12	10	10	12
Typical thickness mm	0.1–1.0	0.1 (bond coat)	0.75–2.5	0.75–2.5
Cost, £/kg	22	44	4	8
Special features	Used for repair work especially on journals where it gives good wear resistant low friction surface.	Used mainly as bond coat. Can also be used in thicker coatings especially to provide moderate wear resistance at higher temperatures.	Good wear and moderate corrosion resistance. Widely used material for machine components. Can be machined with carbide tools but usually ground. Cheap.	Machines easily to give excellent finish. Gives good soft bearing surface. Used on copper and copper alloy substrates.

The most widely used metal wires and some of the properties of the corresponding coatings are given in Table 10.5. In addition, a range of plain carbon steels is used where the environment is non-corrosive and the wear resistance required is only moderate. A range of austenitic steels is also used for more corrosive conditions.

There is little restriction in shape, size or material of components for wire spraying. However, thicker coatings can be tolerated on plain cylinders than on cylinders containing, for example, keyways or on flat surfaces. Care must be taken to ensure that oil is removed from pores in castings or in previously sprayed material.

Substrates of copper and aluminium and their alloys can be coated but bond strengths are low even when molybdenum and nickel aluminide bond coats are used.

The main advantage of the wire process over powder flame spraying is the relatively low cost of the consumables. The main disadvantage of the process compared with plasma spraying is the restricted number of materials which can be sprayed and the relatively high coating porosity.

Solid rods of ceramic materials (usually alumina) can be fused and atomised in the same way as metallic wires. The rods are usually about 25 cm long, longer rods being readily broken. Coating porosities ($\sim 8\%$) are higher and bond strengths (7 N/mm^2) lower than in the corresponding plasma sprayed coatings. Coatings which are normally sealed with an epoxy or phenolic resin are used in low load abrasion applications such as pump shafts and impellors. The

advantages of the process are the relatively low cost of the equipment and its ease of operation (Norton Rokide process).

A more recent development is the packing of ceramic or other brittle powders in plastic tubes. These tubes are available in diameters from 1.5 to 6.35 mm and in lengths up to 1.50 m. In spraying, the plastic tube burns and the sprayed deposits show no trace of carbon residues. In the case of self fluxing Ni–Cr–B alloys a denser deposit is obtained than with the powder flame gun and thus shrinkage on fusion is reduced. With ceramics the deposits are similar to those obtained from solid rods. Mixtures of materials, such as cermets, can also be sprayed using flexible cords.

(iii) Electric arc processes

(a) Arc wire spraying

The arc wire spraying process is illustrated in Fig.10.1(c). Two wires are continuously fed into the gun in such a way that they converge with a small included angle between them. An arc is struck between the wires which continuously melt as they are fed together into the gun. A jet of atomising gas, usually air, is directed at the arc.

The power rating of this equipment is about 15 kW and this allows very high deposition rates to be achieved ($\sim 12 \text{ kg/h}$ for stainless steel). However, the process is noisier and also gives off considerably more fume than combustion gun wire spraying and is therefore only used where high deposition rates are

Table 10.6 Plasma sprayed coatings (metals)

Coating	Stellite type alloy	Molybdenum alloy	Tribaloy	Nickel-aluminium
Composition, %	Cr 25; Ni 11 W 7.5: C 0.5 Co bal	Mo 75; Cr 4.3 Ni 17.8; B 0.8 Si 1.0	Co 62; Mo 28 Cr 8; Si 2	Ni 95 Al 5
Hardness, HV	350	650 (Mo phase) 850 (Matrix)	600 (Matrix) 1100 (Laves phase)	200
Bond strength, N/mm² (lb/in²)	20 (3000)	23 (3400)		24 (3500)
Density, g/ml	7.3	8.0	7.5	7.4
Porosity, %	<2	~2	2	3
Thickness, mm upper limit	1.8	1.0	1.8	None
Surface finish, μm CLA. As-sprayed	9–11	8–13	8–10	15
Temp limit, °C	850	345	980	900
Approx price, £/kg	40	50	31	22
Special features	Good corrosion and high temperature oxidation resistance. In fretting applications superior to chromium carbide—nickel chromium cermet (Table 9.7) when only one of two surfaces can be coated. (Harder carbide material causes excessive wear on mating surface.)	Stronger, denser and more consistent coating than wire sprayed molybdenum. Low coefficient of friction against steels and cast irons. Good resistance to adhesive wear. Used on some piston rings.	One of series of cobalt and nickel based materials (Triballoy) developed by Du Pont. Good resistance to adhesive wear, low coefficient of friction against steels and super-alloys and good oxidation resistance up to 1000°C	Used as bond coat. Good high temperture oxidation resistance.

Table 10.7 Plasma sprayed coatings (oxides)

Coating	Alumina (Impure)	Alumina-titania	Chromium oxide
Composition, %	Al_2O_3 94 TiO_2 2.5 SiO_2 2.0	Al_2O_3 87 TiO_2 13	Cr_2O_3 98
Particle size, μm	$-45 +5$	$-53 +15$	$-30 +10$
Hardness (HV)	770	850	1300
Bond strength, N/mm^2 (lb/in^2)	7 (1000)	14 (2000)	20 (3000)
Density, g/ml	3.3	3.5	5.0
Porosity, %	1–2	~1	1–2
Thickness, mm, upper limit	0.60	0.50	0.4
typical	0.1–0.25	0.1–0.25	0.1–0.25
Surface finish as sprayed, μm CLA	6–8	6–8	2–3
Cost, £/kg	9	16	27
Special features	Low cost general purpose coating. Can be ground with silicon carbide wheels. High wear and corrosion resistance.	Main advantage is ability to be ground and lapped to smoother surface finish.	Highest wear resistance of oxide coatings. Diamond wheels required for finish grinding.

Table 10.8 Plasma sprayed coatings (carbide cermets)

Coating	Tungsten carbide—12% cobalt	Tungsten carbide—17% cobalt	Chromium carbide—25% nickel-chromium
Composition, %	WC 88 Co 12	WC 83 Co 17	Cr_2C_3 75 Ni 20; Cr 5
Particle size, μm	$-40 +15$	$-53 +10$	$-40 +15$
Hardness, HV			
Macro	670	670	600
Carbide particles	1300	1150	1850
Bond strength, N/mm^2 (lb/in^2)	55 (8000)	~68 (~10,000)	41 (6000)
Density, g/ml	14.5	11.5	6.2
Porosity, %	~1	3	3–4
Thickness, mm, upper limit	0.5	0.4	0.4
typical	0.1–0.2	0.1–0.2	0.1–0.25
Surface finish, μm CLA as-sprayed	6–9	5	5–6
Approx price, £/kg	42	39	27
Special features	Best wear resistance of all plasma sprayed coatings. Max. temperature of use 500°C. Expensive material.	Higher Co content gives coating more resistance to impact. Slightly lower (~20%) abrasive wear resistance than 12% Co material. Max. temperature of use 500°C.	Very good wear resistance in temperature range 500–850°C at which temperatures protective layer of chromium oxide forms. Used to resist high temperature abrasive, fretting and percussive wear. When used in fretting application both surfaces must be coated.

necessary, for example in applying thick coatings of stainless steel and copper to large rolls for the steel and printing industries respectively.

(b) Plasma spraying

Plasma spraying occupies an intermediate position between thermospraying and D-gun coating. The equipment is considerably more expensive than that required for thermospraying, more noise and ultra-violet light are produced during the process but coatings are denser and more strongly bonded to the substrate. Plasma spraying is considerably less expensive and more widely available than the D-gun process.

The tungsten cathode and the surrounding annular copper nozzle which are both water cooled are contained in a gun (Fig.10.1(d)) which can be hand held but is more usually mounted on a mechanical traverse system. With modern equipment the power capacity is around 80 kW, but most plasma spraying is done at lower powers, typically 40 kW, with current and voltage about 500A and 80V respectively.

On passing through the arc, diatomic gases such as

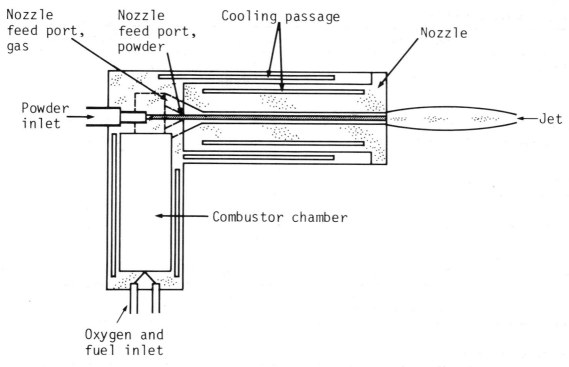

Fig. 10.2   Diagram of 'Jet-Kote' gun[3]

hydrogen and nitrogen are dissociated and all gases are substantially if not completely ionised. The gas temperatures obtained are 4000–8000°K with hydrogen and nitrogen, 15,000°K with argon, and 20,000°K with helium. In the plasma flame nitrogen and hydrogen molecules reform with the evolution of large quantities of heat over a narrow temperature range so that these gases are very effective heating media. However with both nitrogen and hydrogen much higher nozzle wear occurs and this offsets the good thermal properties and lower cost of these gases. The most commonly used arc gas is therefore 90% A/ 10% H₂ which gives adequate heat transfer to most powders without excessive nozzle wear. For very high temperatures a 60% A/40% He mixture is used. Gases issue from the plasma gun at velocities in the range 200–500 m/s, velocities being controlled by nozzle bore and power level.

A number of parameters must be optimised to obtain highest coating density and adhesion. These include, nozzle type, gas pressure and flow rate, power level, powder feed rate, and spray distance. Recommended parameters are given by the various equipment manufacturers.

A large number of powders for plasma spraying have now been developed by the various manufacturers. Many of these are minor variants of main coating types designed to meet specific and limited applications. The variations may be in composition or in particle size. A finer particle size gives a denser coating and also a smoother 'as-sprayed' finish so that the coating can be used without or with a minimum of subsequent grinding. Fine powders can however only be sprayed at a low rate and in the case of oxides and carbides only relatively thin coatings can be applied without loss of bond strength. The particle sizes in the coarser powders are usually in the range

−60 + 15 μm and in the finer powders −12 + 5 μm. The as-sprayed coating thickness limit for the finer oxide and carbide powders is approximately 0.25 mm. Coating thicknesses up to 2.5 mm can be tolerated with the coarsest powders (−75 + 30 μm). Examples of the main coating types are shown in Tables 10.6, 10.7 and 10.8).

Metallic coatings of the Stellite type are used mainly to resist fretting at high temperatures (500–700°C) whilst molybdenum based and Tribaloy materials have good resistance to adhesive wear (scuffing). Tribaloy alloys also have good corrosion and oxidation resistance and are sometimes used below porous oxide coatings to prevent substrate corrosion.

Oxide coatings are hard, wear resistant and the most chemically inert of the plasma sprayed coatings. They are also relatively cheap. However because of their low ductility they cannot be used in any situation involving impact or shock loading.

All carbide based coatings have a substantial metallic content and are therefore tougher and more impact resistant than oxide coatings. Tungsten carbide-cobalt alloys have outstanding wear resistance but start to oxidise above about 500°C. For higher temperature applications chrome carbide-nickel-chromium alloys are used.

Plasma sprayed coatings can be applied to most metals, both ferrous and non-ferrous, and a number of polymeric materials. The only restriction on component size is the size of the spray booth and handling equipment available. With normal spray equipment it is only possible to coat internal surfaces to a depth equal to the diameter of the opening. Moreover the bond strength diminishes when the incident angle is less than 90°C. Purpose built equipment is available which can be used for coating tube bores above 5 cm diameter.

131

During spraying the component temperature seldom rises above 200°C, air or CO_2 cooling being used if necessary. Except on thin (<2 mm) substrates there is little or no specimen distortion and coatings can therefore be applied to finish machined parts.

(iv) Recent developments

Processes which have recently been developed include the high velocity combustion system, originally referred to as 'Jet-Kote', and plasma systems aimed at preventing the oxidation of carbide and metal powders during spraying.

(a) Jet-Kote Process

This process was originally developed by the Browning Corporation but was bought in 1982 by Cabot who improved the design and now market a complete system for $35,000. In this process an oxygen/hydrocarbon (usually propylene) mixture is burned under pressure 0.4–0.6 N/mm^2 (60–90 psi). Gas exhausts from the combustion chamber through four converging channels into a nozzle which can be either 15 or 30 cm long (Fig.10.2). Gas velocities are approximately 1400 m/s, particle velocities are claimed to be in the range 900–1000 m/s, and temperatures up to 2100°C. The gun is light and can be readily operated by hand. Noise is lower in both level and frequency than with plasma.

Because of the high particle velocities coating-substrate adherence is good (>75 N/mm^2) and a hardness of 1275 HV has been measured on tungsten-carbide-12% cobalt coatings which compares favourably with D-gun coatings. Tests are currently being carried out in both Europe and the USA to determine the suitability of Jet-Kote tungsten carbide coatings on the abutting faces of titanium compressor blades. Results, to date, have not been published.

A micrograph of a tungsten carbide-17% cobalt coating applied by a high velocity combustion process similar to Jet-Kote is shown in Fig.10.3 along with a plasma sprayed coating of the same composition. The high velocity combustion process coating has fewer pores, a finer structure and is slightly harder than the plasma sprayed coating. It also appears to be bonded better to the substrate.

The operating cost of spraying one pound of tungsten carbide-12% cobalt is claimed by Cabot[3] to be $4.45 and with a deposit efficiency of 72% this rises to $6.02 per pound deposited. (In the UK the price of tungsten carbide-12% cobalt powder ranges between £30 and £60 per pound depending on the supplier).

(b) Low pressure (or vacuum) plasma spraying (LPPS or VPS)

In the last ten years several low pressure plasma systems have been developed (Fig.10.4). By adjusting the throttle valve between the pump and the chamber the operating pressure can be controlled from atmospheric down to about 40 torr. The chamber is first flushed with argon and during operation the plasma gas maintains an inert atmosphere. There are a number of advantages over "in air" plasma spraying.

(a)

(b)

Fig. 10.3 Micrographs of WC–17%Co applied by (a) high velocity combustion process (HVCP) × 320; and (b) 40 kW plasma × 320. (HVCP coating by courtesy Mr M Donovan, BAJ-Vickers.)

132

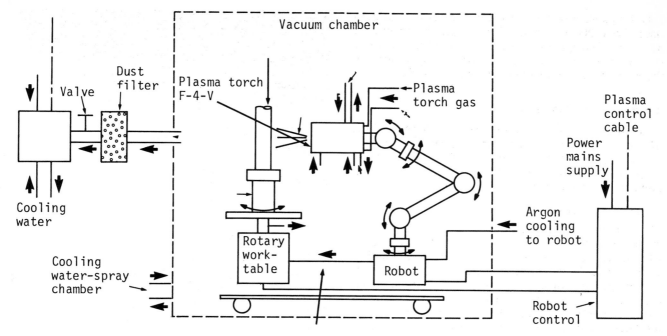

Fig. 10.4   Low pressure (or vacuum) plasma spraying (LPPS or VPS).

The reduced oxidation of powders during spraying and virtual elimination of oxide inclusions.

The ability to use higher substrate temperatures (up to 1000°C) without oxidation, thus reducing stress in coatings and allowing the production of thicker (up to 2.5 mm) and more adherent coatings.

A longer plasma flame[4] so that gun-substrate distance need not be so accurately controlled (Fig.10.5). The angle of impingement of particles on the substrate is also less critical.

A higher deposit efficiency.

Substrate surfaces can be cleaned without prior grit blasting by using an auxiliary arc,

Higher particle velocities and therefore, for a given gun power, denser and more adherent coatings are obtained.

The main disadvantage is the high cost. A production unit with the necessary component manipulation equipment costs approximately £400,000.

Fig. 10.5   Variation of centre line plasma temperature with spray distance at different pressures.[4]

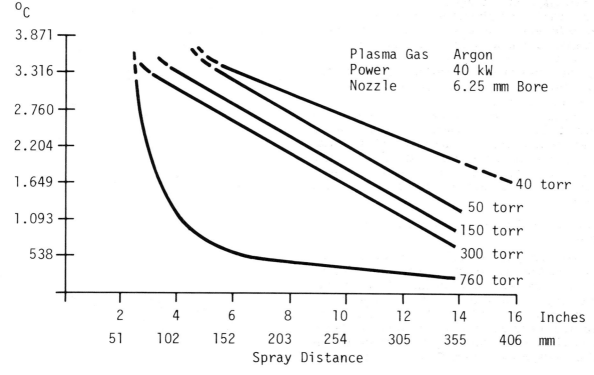

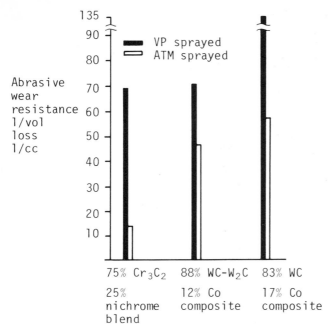

Fig. 10.6 Abrasive wear resistance of carbide cermets deposited by conventional and low pressure plasma.[4]

Suppliers of equipment include:

Electro Plasma, Inc.   Irvine, California
Plasma Technik AG   Wohlen, Switzerland
Metco, Inc.   Westbury, L.I. New York

The main use of the system so far has been in the application of MCrAlY oxidation resistant coatings to aircraft turbine blades. However by preventing oxidation of WC and Cr_3C_2, coatings of WC-Co[4] and Cr_3C_2-Ni/Cr[5] cermets with greater wear resistance are obtained. This is illustrated in Table 10.9 and Fig. 10.6.

In the 12% and 17% cobalt-tungsten carbide cermets the carbon in the LPPS coating was 80 and 91% respectively of that in the original powder compared with 52% and 84% for conventional spraying.

(c) Shrouded plasma spraying

Oxidation during plasma spraying can also be reduced by surrounding the spray cone by an annulus of inert gas. A system patented by Metco Inc.[6] is illustrated in Fig. 10.7. The inert gas which is preferably heated to about 400°C is introduced at the exit end of the shroud and flows in the opposite direction to the plasma gas. Systems have also been developed by Union Carbide[7] and by United Technologies Inc. (Gatorgard). Various systems have been studied at Eindhoven University, the Netherlands.[8] Results in the Metco patent show that when spraying a cobalt base alloy the oxide content is reduced from 16% to 2% using a shroud of nitrogen heated to 400°C and flowing at 1800 cu.ft/hr. Although high quality MCrAlY coatings can be obtained[9] indications are that this is not done so consistently as with LPPS.[10]

Microstructures of MCrAlY coatings obtained by various processes[10] are shown in Fig. 10.8.

(v) Surface finishing operations

In many applications the surface finish of sprayed coatings must be modified before the component is put into service. In the case of ceramic coatings for use in the textile industry the coefficient of friction between synthetic yarn and the contacting surface is especially important. The most frequently used techniques are:

(a) Abrasion

A series of silicon carbide papers ranging from 80 to 260 mesh are applied wet to the surface. Surface finishes in the range 0.6–0.9 μm CLA can be obtained. The friction coefficient with synthetic yarn is relatively high.

(b) Brushing

This is done using cloth brushes impregnated with either alumina or silicon carbide. The surface obtained is around 3 μm CLA and is referred to as nodular or orange peel. The friction coefficient against synthetic yarns is low.

(c) Grinding

Grinding is best done using resin bonded diamond wheels but for reasons of economy silicon carbide can be used on alumina and alumina-titania coatings. Diamond wheels are essential for tungsten carbide-cobalt and chromium oxide coatings. Using progressively finer diamond wheels a surface finish of around 0.3 μm CLA can be obtained. With silicon carbide wheels a surface finish of about 0.8 μm CLA is obtainable. These surfaces have relatively high coefficients of friction against synthetic fibres.

(d) Lapping

Lapping is done using cast iron, copper and hardwood laps in conjunction with fine diamond powder contained in a paste. The size of powder is determined by the required surface finish but is usually in the

Table 10.9 Comparative wear resistance of Cr_3C_2/Ni-Cr coatings applied by conventional and low pressure plasma[5]

	75 w/o Cr_3C_2 25 w/o Ni-Cr (conventional plasma)	50 w/o Cr_3C_2 50 w/o Ni-Cr (conventional plasma)	50 w/o Cr_3C_2 50 w/o Ni-Cr LPPS
Powder type*	Blend	Composite	Composite
Cross-section hardness HV	600	620	860
Surface texture (μm)	1–10	8–10	8–10
Abrasive slurry wear resistance	1	2.8	4.3

*Blend—mixture of powders.
Composite—Cr_3C_2 particles coated with Ni-Cr layer.

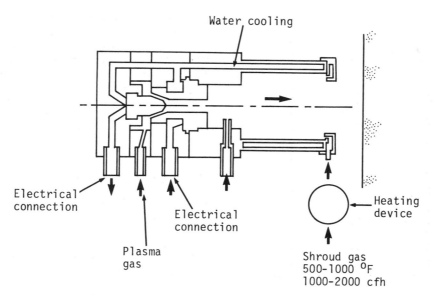

Fig. 10.7   Diagram of shrouded plasma.[6]

Fig. 10.8   Microstructure of MCrAlY coatings applied by various processes.[10] (-X200) (Courtesy of Mr. J. J. Hermanek, Alloy Metals Inc.)

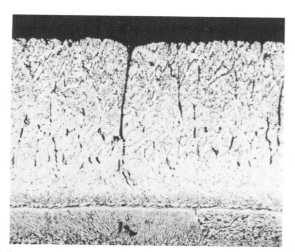

(a) **By E.B. vacuum evaporator.**
    (deep intercrystalline voids)

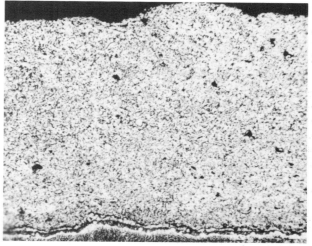

(b) **By vacuum plasma spraying.**
    (relatively free of oxide and voids)

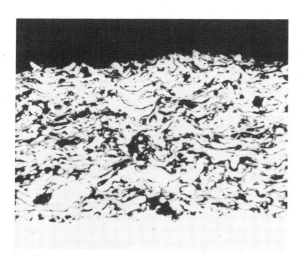

(c) **By normal 'in-air' plasma.**
    (relatively high level of oxide and voids)

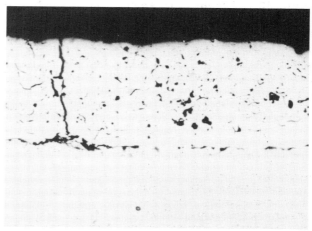

(d) **By shrouded plasma.**
    (Some coating cracks and voids)

Table 10.10 Costs of some sprayed coatings

Coating	Material cost £/kg	Process	Thickness mm	Coating Cost £
13% Cr steel	4.5	Wirespray	0.7	25
Molybdenum	31	Wirespray	0.7	50
Al-bronze	12	Wirespray	0.7	32
Ni-Mo-Al	42	Thermospray	0.7	67
Al_2O_3/3% TiO_2	6.0	Plasma	0.3	27
Cr_2O_3	24	Plasma	0.3	45
WC/17% Co	39	Plasma	0.3	60

Coating shaft 50 mm dia × 500 mm long (785 cm^2) to thickness stated

range 5–20 μm. Surface finishes down to 0.05–0.10 μm CLA can be obtained. Such surfaces have a high coefficient of friction against synthetic yarn.

(vi) Coating costs

The costs of some sprayed coatings, which are believed typical of those prevailing in the UK at the present time, are given in Table 10.10. Plasma coating and thermospraying costs are in the same range, the reduction in coating material usage compensating for higher equipment depreciation. D-gun coatings are approximately twice the price of the corresponding plasma coatings.

(vii) Quality control[11]

Methods of quality control can be grouped under:

(a) Process control;
(b) Destructive testing;
(c) Non-destructive testing.

(a) is the most effective; (b) gives useful information with the reservation that the coupon used may not have the same cross-section nor have been prepared in exactly the same manner as the component; apart from dimensional checks the usefulness of (c) has still to be shown.

(a) Process control

The parameters which require to be controlled to produce optimum coatings can be grouped as follows:

Substrate properties:

> hardness;
> degree of surface roughening;
> surface cleanliness.

Raw materials:

> compressed air —water and oil content;
> plasma gases —oxygen content;
> grit blast medium—size range, purity;
> coating material —composition, size range, water content.

Coating parameters:

> power input;
> gas flow rates;
> powder feed rates;
> gun-substrate distance;
> degree of preheat.

The deposit efficiency which is the ratio of material adhering to a test plate to that passing through the gun is widely used as an indication of process control. High values of deposit efficiency are obviously desirable for economic operation but are also in general a measure of coating quality.

The degree of automatic control of coating parameters is steadily increasing with developments in spraying and component handling equipment. The complete elimination of the operator from the process is the objective of current development programmes.

(b) Destructive testing

Microscopic examination of sections of coating can provide useful information on coating porosity, hardness, coating adhesion and degree of grit embedment at the interface. Careful specimen preparation preferably on automatic equipment is essential for consistent results.

Coating adhesion can be measured using the adhesive pull-off test (ASTM C633-79) but this is usually only done only in development work. A simpler test for routine quality control is that in which an 18 SWG (0.048 in) sheet metal specimen is bent through 90° over a $\frac{1}{2}$ inch diameter roller and the coating surface, which is on the outside of the bend, then inspected for spalling.

(c) Non-destructive testing

Coating thickness is determined by measurements using a micrometer before, during and after the coating process. Allowance must be made for thermal expansion of the substrate. A magnetic pull-off gauge is also used on non-magnetic coatings on magnetic substrates.

Comparative estimates can be made of porosity by polishing small areas of the coating and examining these under a microscope at low magnification.

Other non-destructive tests, eg ultrasonic, eddy current, thermal imaging, acoustic emission, have been used on coatings in development work. Although acoustic emission has given useful information on the cracking of thermal barrier coatings during thermal cycling[12], none of the methods has so far been shown to be suitable for routine quality control.

(viii) Health and safety considerations

The following are the main factors to be considered:

(a) Inhalation of fine dusts and poisonous gases

In all spraying processes some material is evaporated, the amount depending on the process temperature.

On cooling, this material forms very fine particles which if inhaled can be injurious. Processes should therefore be carried out in booths operated under strong suction, fine particles being entrapped in a water wall curtain at the rear of the booth. In plasma spraying the poisonous gases ozone and nitrous oxide can also be produced in the region of the arc and these are also exhausted through the booth.

(b) Noise

All spraying processes produce noise, the intensity increasing with gas velocity. With thermospray processes and arc wire spraying the use of ear protectors is desirable; with plasma spraying, where noise levels can reach 120 dB, it is mandatory. Detonation gun coating, where noise levels reach 150 dB, must be carried out by remote control in sound-proofed cubicles.

(c) Ultraviolet radiation

In arc and plasma spraying protection for the eyes and skin is necessary against the high intensity UV radiation produced.

An increasing amount of plasma spraying is now being carried out automatically in fully enclosed booths so that the operator is completely protected from fume, noise and radiation.

3. Bonding of coatings to substrates

No completely satisfactory explanation of the bonds formed between sprayed coatings and substrates has as yet been put forward. Mechanical interlocking is generally believed to be the predominant mechanism although in spraying metals and cermets metal to metal bonding is not ruled out.

To obtain good bonding a clean roughened surface, usually produced by grit blasting, is essential. However, grit blasting does not produce many re-entrant cavities, which are necessary for mechanical interlocking. Moreover it is also established that the efficacy of grit blasting diminishes with time and that after about two hours exposure to air a grit blasted surface has poor bonding properties, although the surface roughness has not diminished in this time.

Grit blasting, in addition to roughening the surface and increasing its area, plastically deforms and cleans the surface, thus leaving it in a chemically active state suitable for metallic or chemical bonding. Such chemical activity would decrease with time which, if these mechanisms were operative, would explain the effect of exposure on grit blasted surfaces.

It has also been suggested that because the asperities on a grit blasted surface are to some extent thermally isolated from the remaining surface they will reach a higher temperature during spraying, and thus more readily form metallic bonds with the impacting particles of sprayed materials.[13]

The action of bond coats such as molybdenum and nickel-aluminium mixtures is also obscure, although in the case of nickel-aluminium it is generally believed that the exothermic reaction which occurs between nickel and aluminium to form the nickel aluminides NiAl and Ni_3Al, raises the temperature at the particle substrate boundary and thus aids bonding.[14]

The method of measuring the tensile bond strength of sprayed coatings which is most frequently used is specified in ASTM C633-79. In this the coating to be tested is sprayed on to the suitably prepared end of a bar of the substrate material and another bar of the same material is attached to the free surface of the sprayed coating by means of one of several specified adhesives. After curing, a tensile load is applied to the test piece using self-aligning fixtures until rupture occurs. It is emphasised in the specification that there is danger of the adhesive penetrating the porous coatings to the interface and thus giving an erroneously high value of the bond strength. For this reason it is recommended that the method is only used on coatings thicker than 0.4 mm and that certain specified adhesives, preferably cold curing in the case of ceramic coatings, with bond strengths in the range 28–55 N/mm² (4000–8000 psi) be used. Values of bond strength measured by this method on coatings considerably thinner than this and using hot curing resins with bond strengths of up to 80 N/mm² (12,000 psi) are nevertheless quoted in the literature on plasma spraying without mention being made of any precautions taken to prevent adhesive penetration. For very high bond strength measurements a brazed joint can be used between the coating and the test bar.

In summary it can be said that neither the method used to measure nor the theories advanced to explain the adherence of sprayed coatings to substrates are satisfactory. Nevertheless the following facts appear to be established.

A freshly roughened surface (3–8 μm CLA) is essential for good bonding between the substrate and most coating materials.

The self bonding materials, molybdenum and nickel-aluminium will bond to relatively smooth surfaces (< 3 μm CLA).

Bond strengths decrease with increasing coating thickness;[15] this is illustrated in Fig. 10.9 for titanium carbide steel cermet coatings on a steel substrate.

The finer the particle size of the sprayed material the greater the decrease in bond strength with coating thickness.

Bond strengths increase with increasing particle velocity.

There is, at least in some systems, an optimum substrate pre-heat temperature for maximum bond strength;[15] this is illustrated in Fig. 10.10 again for titanium carbide-steel cermet coatings on a steel substrate.

The importance of bond strength obviously depends on the duty to which the coated component is subjected. Where impact forces are present then high bond strengths are essential. The best known example is the abutment faces on aero-engine compressor blades where only the high bond strengths > 140 N/mm²

(>20,000 psi) of the tungsten carbide coatings obtained by the D-gun process have proved adequate. However, in many applications where the components are subjected to low load abrasion then high bond strengths are unnecessary. For example, 'Rokide' type coatings with a bond strength of ~7 N/mm (1000 psi) have for many years been used to resist abrasive wear in pumps.

4. Effect of coatings on fatigue strength

A number of investigations of the effect of sprayed coatings on the high cycle fatigue properties of the substrate have been carried out[16,17,18]. Some aspects of the results are contradictory but some tentative conclusions can be drawn.

Grit blasting carbon and low alloy steels has little overall effect on fatigue strength, the deleterious effects of surface roughening apparently being compensated by the beneficial effects of the compressive stresses introduced. Shot peening prior to grit blasting raises the fatigue strength. It is probable that in higher strength material the deleterious effects of grit blasting will predominate.

In high heat input processes, such as thermospraying and arc wire spraying, martensite may be formed on the surface of air hardenable steels with a consequent reduction in fatigue strength.

The application of hot coating material to a relatively cold substrate leads to tensile stresses at the coating-substrate interface. Their magnitude depends on the thermal gradients produced during coating which in turn depend on the rate of deposition and the thickness of coating. These tensile stresses together with the defects present in sprayed coatings frequently lead to a reduction in high cycle fatigue strength.

5. Wear resistance of coatings

Some information on the wear resistance of various sprayed coatings are given in Fig. 10.11 and Table 10.11. Fig. 10.11 shows the abrasive wear resistance of various plasma sprayed coatings relative to hard chromium plate.[19] Carbide and oxide coatings are more whereas the metal coatings are less abrasion resistant than chromium plate. On the same scale plasma sprayed tungsten carbide-12% cobalt would be approximately 270.

Table 10.11 shows the volume of material removed from a spherical surface of the test material when rubbing against a flat plate of the abutting material.[20] These results are relevant to applications in aero-engines and values of 1–10 are considered good, and 10–30 acceptable. The results again show the outstanding abrasive wear resistance of tungsten carbide-cobalt applied by both D-gun and plasma gun and also the good wear resistance of other less expensive materials such as wire-sprayed 13% Cr steel

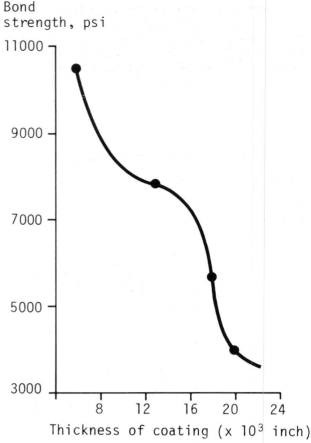

Fig. 10.9   Effect of coating thickness on bond strength[15] (titanium carbide-steel cermet coatings).

Fig. 10.10   Effect of substrate preheat temperature on bond strength.[15] (titanium carbide-steel cermet coatings).

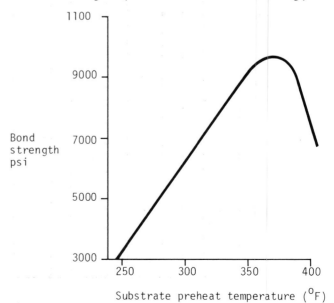

when rubbing against itself. The softer high chromium materials show a very high wear rate at the lower temperatures, presumably due to adhesive wear, which diminishes at higher temperatures with the formation of chromium oxide films which prevent contact between the metal surfaces.

References

[1] TUCKER, R C. Plasma and detonation gun deposition techniques and coating properties. *Deposition technologies for films and coatings*. ed Bunshah, R F. Noyes Publications, NJ 1982.

Fig. 10.11 Abrasive wear resistance of various sprayed coatings relative to hard chrome plate.[19]

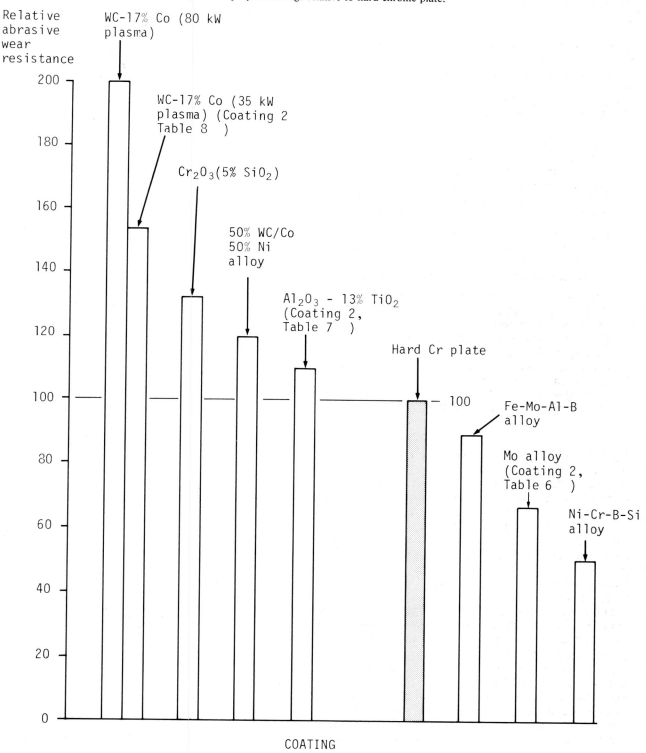

Table 10.11 Dry rubbing wear test results[20]

Material under test	Abutting material	Material removed, μin^3, at following temperatures, °C						
		20	100	200	300	400	500	600
Structural materials								
12% Cr steel	12% Cr steel	230	111	6.35	4.65	1.71	3.5	11.35
12% Cr steel	Electro plated chromium	69.8	33	10.8	2.85	0.35	0.41	—
12% Cr steel	Molybdenum	37.1	37.6	36	11.7	5.0	—	—
12% Cr steel	Metco 404	—	—	9.4	6.65	0.7	1.19	9.4
18/8 stainless steel	18/8 stainless steel	122	31.2	15.8	11.7	7.0	4.35	6.7
18/8 stainless steel	Electro plated chromium	35	21.4	17.9	12.8	24.4	2.1	—
18/8 stainless steel	Molybdenum	77	55	27	16.4	12.8	—	—
18/8 stainless steel	Metcoloy 2	—	11.6	7.0	4.3	0.2	0.67	1.23
18/8 stainless steel	Metco 404	12.6	11.6	8.0	3.5	2.18	5.4	13.5
Nimonic 80A	Nimonic 80A	123	117	98	103	117	116	133
Nimonic 80A	Electro plated chromium	25.6	24	19.8	5.4	44	20	26
Nimonic 80A	Molybdenum	61	67.8	—	—	—	—	—
Nimonic 80A	Metcoloy 2	17.5	20.4	7.3	2.1	0.34	0.1	0.16
Nimonic 80A	Metco 404	14	20.7	9.9	6.65	18.1	30.1	—
Electro plated materials								
Chromium	Electro plated chromium	7.6	30.3	19.1	16.4	3.25	1.1	0.57
Chromium	12% Cr steel	3.9	8.0	4.1	1.31	0.2	0.08	—
Chromium	18/8 stainless steel	7.2	4.55	5.3	4.8	8.2	—	—
Chromium	Nimonic 80A	1.8	9.5	20.4	4.4	39.4	37.6	14.3
Oxy-acetylene materials								
Molybdenum	Molybdenum	13.5	16.7	11.3	2.48	—	—	—
Molybdenum	12% Cr steel	7.6	35.5	17.2	6.75	4.05	—	—
Molybdenum	18/8 stainless steel	54.0	26.8	23.1	8.5	7.1	—	—
Molybdenum	Nimonic 80A	16.6	20.0	—	30.8	—	—	—
Metcoloy 2	Metcoloy 2	2.53	7.1	8.52	6.65	0.31	0.11	0.18
Metcoloy 2	18/8 stainless steel	—	0.05	11.6	3.9	0.5	0.41	1.61
Metcoloy 2	Nimonic 80A	9.2	30.1	22.9	3.5	3.35	0.44	0.99
Metco 404	Metco 404	11.3	14.2	7.5	2.07	6.15	6.55	3.25
Metco 404	12% steel	—	—	14	4.4	0.92	0.64	8.2
Metco 404	18/8 stainless steel	164	91	11.9	7.8	1.28	2.0	8.4
Metco 404	Nimonic 80A	65	51.5	22.0	11.2	13.5	30.1	—
Plasma sprayed materials								
XP1110	XP1110	0.28	0.28	0.41	2.0	1.51	0.48	—
XP1110—AS	XP1110—AS	5.3	9.9	9.5	6.65	—	—	—
XP1139	XP1139	0.05	0.04	5.95	4.75	6.37	8.95	12.5
XP1139—AS	XP1139—AS	5.0	6.8	11.8	7.5	19.2	22.8	35.4
Stellundum 45 AS	Stellundum 45 AS	3.62	4.25	12.25	19.4	18.7	26.0	—
Stellundum 52H	Stellundum 52H	0.01	0.01	0.01	0.27	0.25	0.25	—
Stellundum 52H AS	Stellundum 52H AS	0.79	0.67	Nil	0.30	Nil	Nil	—
D-gun materials								
LW1	LW1	Nil	Nil	Nil	Nil	Nil	Nil	0.08
LW1—AS	LW1—AS	0.34	0.2	0.32	0.41	0.48	0.31	0.6
LW1N30—AS	LW1N30—AS	0.68	0.70	0.92	0.79	1.05	0.54	—

All tests conducted with a ground surface finish unless stated otherwise ie AS—as sprayed finish
LW 1 WC/9% Co (D-gun) ⎫ Union Carbide
LW1N30 WC/13% Co (D-gun) ⎭
Metco 404 80% Ni/20% Al (powder)
Metcoloy 2 13% Cr (wire)
XP1110 WC/12% Co
XP1139 50% WC/Co/50% Ni based self fluxing alloy ⎫ Metco
Stellundum 45 50% WC/Co/50% Ni based self fluxing alloy ⎫ Cabot Corp.
Stellundum 52H WC/16% Co

2 BROWNING ENGINEERING CORPORATION. US Patent 4,416,421, 1981.

3 WHITFIELD, R W, APPELT, J M, MANNO, D E and SOMERVILLE, D A. Supersonic jet spraying; process; equipment and coating characteristics. *International Conference on Metallurgical Coatings*, American Vacuum Society, San Diego, 1984.

4 WOLF, P C and LONGO, F N. Vacuum plasma spray process and coatings. *9th International Conference on Thermal Spray Coatings*, The Hague, 1980.

5 REARDON, J D, MIGNOGNA, R and LONGO, F N. Plasma and vacuum plasma sprayed chromium carbide composite coatings. *Thin Solid Films*, Vol 83, (3), p 545, 1981.

6 METCO INC, British Patent 1,597,558. 1981.

7 UNION CARBIDE CORP. US Patent No 3,470,347. 1968.

8 HOUBEN, J M. Problems encountered in the development of locally shielded plasma spray devices. *9th International Conference on Thermal Spraying*, 1980.

9 TUCKER, R C, TAYLOR, T A and WEATHERLY, M H. Plasma deposited MCrAly air foil and zirconia thermal barrier coatings. *3rd Conference on Gas Turbine Materials in Marine Environment*, Bath, 1976.

10 HERMANEK, F J. Thermal spray coatings—the process and its evolution. *Amer. Electroplaters Soc.*, San Francisco meeting, June 1982.

11 PUTZIER, U. ZICKLER, H and KOCH, F. Quality control with thermal spraying. *9th International Conference on Thermal Spraying*, The Hague, 1980.

12 SHANKAR, N R, BEMDT, C C and HERMAN, H. Structural integrity of thermal barrier coatings by acoustic emission studies. *10th International Conference on Thermal Spraying*, Essen, 1983.

13 MOSS, A R and YOUNG, W J. Arc plasma spraying. p 287. *Science and Technology of Surface Coating*. Academic Press, 1974.

14 INGHAM, H S. Bonding of flame sprayed Ni-Al. *Jnl Vac. Sci. Tech..* Vol 12, (4), 1975.

15 WOLOSIN, S and COHEN, S. Development of a wear resistant coating for the rotary engine. *Proc. 8th International Conference on Thermal Spraying*, Miami, p 45, 1976.

16 MERINGOLO, J, SILVAGGI, L and MOGUL, J. Effect of plasma spray coatings on fatigue properties on H-11 steel and 2024 aluminium. *6th International Conference on Thermal Spraying*, Paris, September 1970.

17 KESHVARZI, A, and REITER, H, Effect of flame sprayed coatings on the fatigue behaviour of high strength steels. *10th International Conference on Thermal Spraying*, Essen, 1983.

18 BERTRAM, N and SCHEMMER, M, Fatigue behaviour of thermally coated steels. *10th International Conference on Thermal Spraying*, Essen, 1983.

19 LONGO, F N, Plasma and flame sprayed coatings satisfy hard chrome plate applications. *8th International Conference on Thermal Spraying*, Miami, 1976.

20 TRSEK, A, The use of thermal sprayed coatings for extended life of aircraft jet engine parts. *8th International Conference on Thermal Spraying*, Miami, 1976.

CHAPTER 11
Weld hardfacing

1. Introduction

The deposition of hard surfaces by welding is usually referred to as hardfacing. The surfaces produced in this way are relatively thick (up to several millimetres) and the process is generally used where large amounts of wear can be tolerated before retreatment of the surface.

This is a process in which a suitably alloyed metal is melted and fused to a metal surface, to provide a hard wear-resistant coating. The necessary heat may be obtained from an oxy-acetylene flame or electric arc, the hard coating material being supplied in the form of a filler rod, wire or consumable electrode, or alternatively as a powder or water-based paste.

Hardfacing is essentially a welding process and, as such, the methods are largely those used in welding practice, developments in which have had considerable influence on hardfacing techniques. This chapter is concerned with those techniques and consumables that have special relevance to hardfacing operations, a broader coverage of welding technology clearly being outside the scope of this book.

Details of the conventional welding processes used in depositing hardfacing can be found in numerous text and reference books.[1,2,3]

Hardfacing is widely used in the repair of worn components, as well as in the manufacture of new equipment where to make whole components from a wear resisting alloy would be impractical or too costly. It is especially useful in situations where lubrication is absent or inadequate, eg in the handling of dry, abrasive materials. The coatings, which may range in thickness from 1 mm upwards, are generally applied only to those areas in which maximum exposure to wear is encountered.

The process is widely applied for the protection of earth and rock engaging equipment used in agriculture, mining, oil well drilling and civil engineering, when, in the majority of cases, the deposits perform satisfactorily in the as-welded condition. In such cases re-building operations can, if necessary, be carried out in the field using portable equipment. Heavy duty knives and guillotine blades on the other hand require finishing to size; this must be done by machining with high speed steel or carbide-tipped tools, or by grinding. If the hardfacing can be done indoors a wide range of automatic equipment is available.

2. System selection

Selection of the most appropriate system, ie material and deposition process, should be made as the outcome of the following sequential steps:

— determine the mode of wear (Chapter 1)

— choose an appropriate hardfacing material taking into consideration:
  (a) the mode of wear, whether sliding, abrasive or impact, and whether toughness or hardness is needed
  (b) the deposit thickness needed to give satisfactory life and whether dilution by the substrate is likely to demand thick coatings
  (c) the temperature of operation
  (d) whether the environment is likely to be corrosive,
  and
  (e) the cost/kg of candidate hardfacing materials, bearing in mind their density and therefore their covering power.
  Reference to Tables 11.1, 11.2, 11.6[4] and Fig. 11.1[5] will help in (a)–(d) above, Table 11.15 and Fig. 11.15 will help in (e).

— select the most appropriate deposition process taking into consideration:
  (a) the substrate composition (p 151)
  (b) the size, shape and number of components
  (c) the area to be coated and the coating thickness
  (d) whether the work has to be done on site or in a workshop
  and
  (e) the availability of equipment together with appropriate operator skills and supporting services.

— compute the cost of hardfacing the components, bearing in mind that:

  (a) pre-heating may be necessary
  (b) excess material may have to be deposited to give a suitable machining or grinding allowance.
  (c) two or three layers of hardfacing may have to be deposited to overcome the deleterious effects of dilution by the substrate
  (d) the maximum deposition rate for the process may not be achievable if components are thin enough to distort with high rates of heat input or if interpass temperatures are high enough to 'overheat' the substrate

and

(e) post deposition machining or grinding costs, depending on the hardness, must be included.

— carry out welding trials to establish the conditions that will deposit the hardfacing to best advantage both in terms of cost and performance.

— try other hardfacing materials and compare the actual wear in laboratory tests or field trials.

It should be borne in mind that manufacturers of consumables, although understandably biased towards their own products, can often help during the early materials and process selection stages, because of their knowledge of successful use in similar applications.

3. Hardfacing materials

Deposits used to resist abrasive wear generally have a structure containing carbides of chromium, tungsten or boron dispersed in a matrix of iron, cobalt or nickel, those used to resist impact generally have austenitic structures that are capable of work hardening eg austenitic manganese steel. Filler rods or electrodes used to produce deposits differ in form and composition according to the welding process adopted: they may be in the form of solid alloy wires, rods or strips, or of tubes containing carbide-forming elements in the core. The electrodes may be bare or coated with a fluxing agent, when the coating may also contain materials that influence the composition of the deposit. Hardfacings of a given composition may therefore be produced by a variety of welding rod or electrode types, each being specially formulated to compensate for compositional changes occurring during the welding process.

Electrodes, rods and wires are nevertheless most conveniently described according to the structure or composition of the deposit they produce, and on this basis the main categories of hardfacing material have been summarised in Table 11.1. This table also shows the designations adopted for each type in the AWS/ASTM and DIN specifications for hardfacing materials.

In this review the different materials are described under the primary classification headings shown in Table 11.1, a classification which groups together materials having similar wear characteristics. Table 11.2 lists these groups in order of increasing abrasion resistance and decreasing toughness. Fig. 11.1 shows their relative heat and corrosion resistance.

(i) Austenitic manganese steels

Austenitic manganese steel deposits depend for wear resistance on their ability to work harden. As deposited the material is relatively soft (approximately 170 HV), but a hardness up to 550 HV develops on

Table 11.1 Hardfacing materials. Types and specifications

Material type		Specification or grade	
Primary classification	Secondary classification	AWS–ASTM[1] designation	DIN[2] alloy group
Austenitic manganese steels	With 2.75–6.0% Ni With 0.6–1.5% Mo	Fe Mn–A Fe Mn–B }	7
Martensitic and high speed steels	Low carbon martensitic steels High carbon martensitic steels High speed steels	— — Fe–A,B,C	3 6 4
Martensitic, austenitic and high chromium irons	Low chromium (<16% Cr) High chromium (26–32% Cr)	— Fe Cr–A1	— 10
Tungsten carbide composites	Graded according to tungsten carbide content, particle size and matrix material	WC	21

Nickel- and cobalt-base alloys	Nickel base*
	  with up to 35% Mo
	  0.3–0.6% C
	  0.4–0.8% C
	  0.5–1.0% C
		
		—
		Ni Cr–A
		Ni Cr–B
		Ni Cr—C
			
			23
			22–40
			22–50
			22–60

	Cobalt base*
	  0.9–1.4% C
	  1.2–1.7% C
	  2– 3% C
		
		Co Cr—A
		Co Cr—B
		Co Cr—C
			
			20–40
			20–45
			20–50

*The carbon content ranges shown correspond to those for welding rods in AWS 13–70; the DIN alloy groups do not necessarily correspond with these, being based on hardness level rather than on carbon content (eg 20–40 is a cobalt-base deposit with a hardness of HRC 40)

[1]American Welding Society Specifications
A5 13–70 Surfacing welding rods and electrodes
A5 21–70 Composite surfacing rods and electrodes
[2]Deutsche Normen
DIN 8555
Schweisszustzwerkstoffe zum auftrag schweissen

surfaces exposed to heavy impact from rock or metal. These steels should not therefore be used to resist abrasion unaccompanied by heavy impact, under which conditions it may be better to use a martensitic iron.

The austenitic structure of manganese steels is normally produced by water quenching from about 980°C, but hardfacing deposits are usually formulated to ensure that the austenitic structure is obtained on air cooling from the welding temperature. The structure is not, however, wholly stable and slow cooling or re-heating, particularly in the range 400–600°C, can cause the precipitation of manganese-iron carbides, with consequent embrittlement. For this reason manganese steels should be welded with rapid heating and cooling, using as low heat input as possible, by limiting amperage and depositing string beads, or placing the workpiece in a water bath. Because of these limitations oxy-acetylene welding is seldom used. In the event of embrittlement after too much re-heating, toughness may be restored by water quenching from 1010°C.

The deposits generally have compositions in the ranges shown in Table 11.3.

The addition of nickel or molybdenum helps to stabilise the austenite and increase the toughness of the lower carbon types; the molybdenum grade has a somewhat higher yield strength. Some wear resistant grades have 14–17% chromium which gives a high impact resistant steel with high corrosion resistance.

The re-facing of worn carbon steel parts with manganese steel is not generally to be recommended; the transition zone will tend to be low in manganese, develop a martensitic structure, and so give rise to spalling. This difficulty can sometimes be overcome by first 'buttering' the carbon steel part with a layer of austenitic chromium-nickel steel, which blends with the base metal and the overlay, without the formation of a brittle layer.

Deposits of this type are extensively used for building up worn areas on manganese steel railway crossings and frogs, and on a wide variety of components used in earth-moving, agricultural and mining equipment, especially where softer rocks are encountered. The work hardened surfaces are not particularly resistant to abrasion by the harder minerals such as quartz when, in the absence of heavy impact, a martensitic

Table 11.3 Austenitic manganese steels

Type	Composition, %				
	C	Mn	Ni	Mo	Cr
Nickel bearing	0.5–0.9	11–16	2.75–6.0	—	—
Molybdenum bearing	0.5–0.9	11–16	—	0.6–1.5	—
Chromium bearing	0.5–0.9	14	—	—	14–17

or austenitic iron will perform better.

The cost of consumables is £3/kg for plain C/Mn steels; up to £8/kg for nickel or chromium grades.

(ii) Martensitic and high speed steels

These deposits generally have a composition that is balanced so that martensite forms on air cooling, its hardness depending on the carbon and chromium content. They provide the best and most economical surfacing materials for applications involving moderately abrasive conditions combined with medium impact, the abrasion resistance increasing with carbon and chromium content at the expense of ductility and impact resistance. A wide variety of compositions fall within this category: the main types, however, are shown in Table 11.4.

The so-called 'semi-austenitic' varieties are intentionally over-alloyed with carbon, chromium and manganese to yield structures in which some austenite is retained and which consequently exhibit better toughness (they are also more corrosion resistant). In these deposits the austenite is work-hardenable.

The lower carbon grades, up to about 0.2% carbon, are machinable, preferably using tungsten carbide tools; with higher carbon contents grinding is necessary. (These grades may, however, be machined if they are fully annealed and furnace cooled. They must then be re-hardened.)

As a class these materials are especially useful in metal-to-metal wear situations, when they exhibit a low coefficient of friction and the ability to take on a high degree of polish. The lower carbon grades are used for die surfacing, for machinery parts such as shafts, gears and cams, and for unlubricated parts such as articulated track sprockets and rollers. The higher carbon grades are useful for building up worn

Table 11.2 Hardfacing materials—in order of increasing abrasion resistance and decreasing toughness

	Material type	Properties of deposits	Type of duty for which deposits are most appropriate
Increasing abrasion resistance → / Decreasing toughness →	Austenitic manganese steels	Tough, crack resistant and soft. Ability to work harden	High stress, heavy impact
	Martensitic and high speed steels	Good combination of abrasion and impact resistance. Abrasion resistance increases with carbon and chromium content at expense of impact resistance.	Non-lubricated metal-to-metal wear.
	Nickel- and cobalt-base alloys	Wear, corrosion and heat resistant, with good all round strength but low ductility.	Abrasive conditions accompanied by high temperature and/or corrosion.
	Martensitic and high chromium irons	Excellent resistance to wear by most minerals. Brittle.	Highly abrasive conditions.
	Tungsten carbide composites	Extreme resistance to sliding abrasion by hard minerals. Worn surfaces become rough.	Extremely abrasive conditions where extra cost is warranted.

Table 11.4 Martensitic and high speed steels

Type	Composition, %				Hardness*
	C	Cr	Mo	W	
Lower carbon martensitic	0.1–0.7	2.5–5.0	2.0(max)	—	
Higher carbon martensitic	0.7–1.5	5.5–9.5	2.0(max)	—	30–65 HRC
Semi-austenitic	0.7–1.5	11–15	2.0(max)	—	(300–750 (HV)
High speed	0.5–1.0	3.0–5.0	4.0–9.5	1.0–7.0	

*depending on the carbon content and the amount of transformation to martensite.

surfaces prior to finishing with a more abrasion resistant but brittle overlay; they are also used for surfacing pulverising hammers and earth engaging components.

The molybdenum and tungsten content of the high speed steels confers increased resistance to impact and wear at high temperatures. The lower carbon varieties are useful for surfacing hot working tools and for components requiring a measure of toughness; the higher carbon types are more suitable for surfacing edges on cutting and machining tools such as shear blades, reamers, broaches, guides and metal forming dies. Machining of these deposits can only be carried out if they are first annealed; otherwise grinding is necessary. Hardening of the deposits is generally unnecessary, but it is usual to subject them to one or two tempering operations, which improves the hardness and strength.

The prices of consumables vary with the alloy content but are generally within the range:

low carbon martensitic	£2–3/kg
higher carbon martensitic	£3–5/kg
semi austenitic	£5–7/kg
high speed	£14–16/kg

(iii) **Martensitic and austenitic high chromium irons**

The common feature of this group is the high carbon content and thus the dominance of hard carbides in the structure. In the case of the lower chromium alloys these are basically iron carbides and form the matrix material; in the higher chromium alloys they are represented by needle-like crystals of Cr_7C_3 in a matrix of austenite and/or martensite. Typical com-

Fig. 11.1 Relative abrasion, impact, heat and corrosion resistance of the hardfacing alloys.[5]

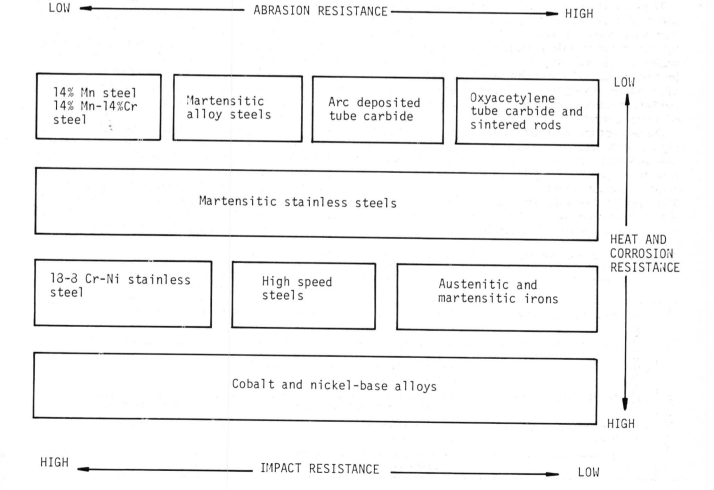

Table 11.5 Martensitic and austenitic high chromium irons

Type	Composition, %				Hardness
	C	Mn	Cr	Mo	
Lower chromium					
Martensitic	3.5—4.0	1.0(max)	4—15	2–4	Carbide matrix
Austenitic	2.0–4.0	2.0–2.5	12–16	Up to 8	1200–1400 HV
Higher chromium					
Martensitic	2.5–4.5	1.0–1.5	26–30	—	Cr_7C_3 particles:
Austenitic	3.0–5.0	Up to 8	26–32	—	up to 1700 HV
					Matrix 400–600 HV

position ranges for the two types are shown in Table 11.5.

The lower chromium martensitic irons are suitable for light impact, but are likely to spall or crack under more arduous conditions. The austenitic low chromium irons are less liable to crack and can be used for medium impact. The higher chromium grades, because of the presence of the harder chromium carbide crystals, exhibit better abrasion resistance, but are brittle and must have adequate base material support; tungsten, molybdenum or vanadium are sometimes added to increase hot hardness, when alloy balance and carbon content are of critical importance.

Materials in this category are particularly resistant to abrasion by wet siliceous sands, clays and minerals. They are therefore widely used for surfacing components in agricultural machinery working in sandy soils (not rock, for which their impact resistance is inadequate), as well as for equipment such as chutes and conveyor screws used in the bulk handling of ores, coke, cement, etc. Deposits have wearing properties that are often comparable with those of the more expensive tungsten carbide composites and the nickel- and cobalt-base alloys.

The cost of consumables varies with chromium content and is generally in the range £5–10/kg.

(iv) Cermets

These materials are composites consisting of hard carbide particles embedded in a metal or alloy matrix and are deposited either from solid composite rods or electrodes, or from powder. In either case the size of the carbide particles is graded according to the intended application; large carbide particles are preferred for cutting, finer particles for abrasive wear resistance. Rods or electrodes can be in the form of a metal tube which contains the carbide, the tube melting to form the matrix, or alternatively the carbide can be in a flux added to the outside of a rod. These rods or electrodes can be used with either oxy-acetylene or arc welding processes. Powder feed is usually only practised with the plasma transferred arc process.

(a) Tungsten carbide composites

In rod or electrode form these consist of either (a) low alloy steel tubes containing 40–70% of cast and crushed tungsten carbide particles (the tubes can in some cases be made from high nickel or cobalt alloys), (b) sintered rods made up of 50–80% tungsten carbide in an iron base matrix containing 10% chromium or nickel or (c) mild steel wire thickly coated with a mixture of fine tungsten carbide particles in a nickel, chromium, boron, silicon matrix. These materials are deposited either by oxy-acetylene or by arc welding, appropriate grades are available for each type of deposition.

The carbide particles, which consist of a mixture of WC and W_2C having a hardness of 2100–2500 HV, do not melt during the welding operation. Some solution occurs under the high temperatures encountered in arc welding and arc currents should be minimised as far as possible if maximum resistance to abrasion is to be secured. For this reason oxy-acetylene is generally the preferred deposition method.

The matrix composition depends on the amount of solution that has taken place and matrix properties may range from those of the air hardened tool steels to cast irons containing secondary tungsten-iron carbides. The more tungsten in the matrix, the better the hot hardness; operating temperatures above about 650°C are however, to be avoided, as the carbides then tend to oxidise. The addition of cobalt to the granules improves the toughness of the matrix, but is accompanied by some loss in abrasion resistance. For certain mining and drilling applications the carbide particles are bonded in a brass or nickel-silver matrix.

The hardness of quartz and most siliceous minerals falls between that of the matrix and the carbide particles, which means that when abraded by these materials, the deposits are selectively worn away, leaving the tungsten carbide particles standing out in relief. Clearly therefore the size of the particles is of importance, and even with fine grades, worn surfaces are rough and cutting tools develop a serrated edge.

These overlays are extensively used to armour the cutting teeth of rock drills and the wearing surfaces of mining, quarrying and earth moving equipment, where the deposits can be used without the need for any machining. The deposits are non-machinable and difficult to grind; they are therefore seldom used in metal-to-metal wear situations.

In powder form tungsten carbide cermets can be deposited by the plasma transferred arc process. The particles used consist of tungsten carbide with either added cobalt, iron/chromium, nickel/chromium/boron or cobalt/chromium/tungsten.

The cost of consumables varies between £30 and £60/kg.

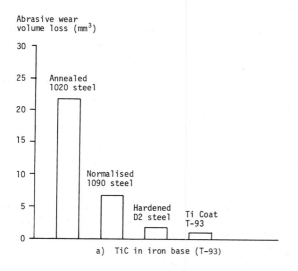

a) TiC in iron base (T-93)

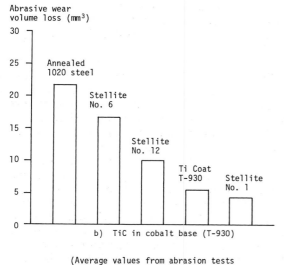

b) TiC in cobalt base (T-930)

(Average values from abrasion tests wet sand/rubber wheel tests)

Fig. 11.2 Comparison of abrasion resistance of titanium carbide cermets with other materials.[6]

(b) Vanadium carbide composites

Vanadium carbide is slightly harder than tungsten carbide (2600 HV compared with 2400 HV) and is much less dense (5.7 g/cm^3 compared with 15.6 g/cm^3 for WC) and for a given weight of electrode or rod will give up to 35% greater coverage. It is claimed that because of the lower density there is less tendency for the carbide to segregate in the iron based matrix. Although the cermet is used in place of tungsten carbide in the less arduous applications, there has been no data published yet about its relative performance.

The cost of consumables appears to be similar to tungsten carbide on a weight basis.

(c) Titanium carbide composites

The high hardness of titanium carbide (3200 HV compared with 2400 HV for WC) has not been utilised in hard facings until recently because of its tendency to oxidise during deposition. Like vanadium carbide it has a very much lower density than tungsten carbide and therefore gives more coverage for a given weight. The plasma transferred arc process has allowed it to

be deposited in powder form in conjunction with iron, cobalt or nickel base matrices. It is claimed that a more dispersed phase is possible than by oxy-acetylene or conventional arc processes. The process is best suited for depositing relatively thin layers on precision parts, where a smooth surface is required.

Comparative wear data with other hardfacing materials is provided by the manufacturer (Fig. 11.2).[5]

The cost of consumables is about £55/kg for an iron base and £70/kg for a cobalt base powder.

(v) Nickel and cobalt-base alloys

These hardfacing materials are widely used where abrasion is accompanied by high temperatures or corrosion. With a corrosion resistance and electrode potential very similar to AISI 316 stainless steel, they are especially suitable for hardfacing stainless steel equipment. Both the cobalt- and nickel-base series contain high percentages of chromium, depending for their abrasion resistance on the presence of chromium carbides and, to a lesser extent on tungsten and/or molybdenum carbides. These carbides have a hardness very similar to those in the austenitic and martensitic irons, which are much cheaper and equally satisfactory for surfacing parts not exposed to high temperatures or corrosion. Both the nickel- and cobalt-based alloys are suitable for metal-to-metal wear situations, when they take a high polish, have a high resistance to galling and a low coefficient of friction. Generally speaking the cobalt-base alloys (the Stellites) are more resistant to temperatures above 650°C than are the nickel-base alloys. Hot hardness values for some typical alloy compositions are given in Table 11.6.[4]

Owing to their high hardness most of these alloys were originally available only as rigid cast rods of restricted length, so that deposition was not possible using high speed processes. A great many of the alloys are, however, now available in granular form, for deposition using powder processes. For automatic oxy-acetylene, TIG or submerged arc welding many of the cobalt alloys are additionally available in the form of continuous coiled wire, consisting of a ductile cobalt strip formed into a lapped tube containing a core of powdered chromium and tungsten. A recent development is the introduction of continuously cast Stellite rods that are free from porosity and inclusions.

In the case of the cobalt alloys, variations in carbon, tungsten and molybdenum content provide hardnesses ranging from about 30HRC to 60HRC. A large number of proprietary electrode wires and powders are available, the compositions of which generally fall within one or other of the grades listed in Table 11.7. The deposits do not respond to heat treatment and are generally used in the as-deposited condition. They are not readily machined, although in the case of the low carbon grade this may be done using tungsten carbide tools. With the higher carbon grades, finishing must be carried out by grinding. Increasing hardness is accompanied by increasing brittleness, so that the higher carbon grades require adequate base metal support.

During the late 1970s the price of cobalt rose steeply

(from \$8.8/kg in 1975 to \$55/kg in 1979) which stimulated a search for hardfacing compositions that gave the properties of established high cobalt alloys but had reduced cobalt contents or none at all. None of these have properties equalling the best high cobalt alloys and now that the price of cobalt is comparatively stable at \$24.2/kg many of these 'substitute' alloys have been withdrawn.

The most important group of nickel-base hardfacing alloys are those containing boron and silicon, deposits of which rely for their wear resistance on the presence of massive borides and carbides, notably of chromium. These alloys, which are characterized by a relatively low melting point, flow easily and bond at about 1000°C. Very thin, smooth deposits are therefore obtainable with this type of material especially when using the oxy-acetylene powder welding, spray-fuse or plasma arc welding processes. The presence of boron may result in some loss of high temperature resistance, due to the formation of low melting point constitiuents. Compositions usually fall within the ranges shown in Table 11.8.

Another group of nickel alloys, developed primarily for their resistance to corrosion in the presence of mineral acids, contain substantial quantities of molybdenum as well as chromium. A typical composition and hardness are shown in Table 11.9.

The hot wearing properties of this alloy are good, particularly in some severely corrosive media. Its wearing properties depend, to a large extent, on the presence of intermetallic compounds which play the same role as the carbides in the cobalt-base alloys. It is used for pump and valve parts in the chemical industry, and also for hardfacing drop forging dies, when it work hardens under the influence of heavy impact.

A series of alloys registered under the name of Tribaloy have an even higher molybdenum content. Based on either cobalt or nickel, their nominal compositions are as shown in Table 11.10.

These alloys, containing less than 0.08% carbon, exhibit particularly good sliding wear resistance under conditions of poor lubrication, properties which results from a structure consisting of a hard intermet-

Table 11.6 Hot hardness of plasma transferred arc (PTA) surfacing powder deposits[4]

Hardfacing material		Hardness (HV)					Type
		RT	800°F 427°C	1000°F 537°C	1200°F 649°C	1400°F 760°C	
Tribaloy	T400	—	605	585	585	335	Co base alloy
	T800	—	555	555	500	295	
Stellite	1	—	510	465	390	230	Conventional Co base alloy
	6	390/450	300	275	260	185	
	12	—	345	325	285	245	
Haynes	711	—	380	335	300	215	Substitute low Co, Ni base alloy (11–12%Co)
	716	—	290	270	235	180	
Hastelloy	C	195	190	185	170	145	Ni base alloy with 2.5%Co
Deloro	60	585/680	555	440	250	115	Ni base alloy
	40	525	365	270	195	80	
Delcrome	90	—	350	255	130	95	Fe base alloy

Table 11.7 Cobalt-base alloys

Type	Composition, %				Hardness		Typical Applications
	C	Cr	W	Co	HRC	HV	
Low carbon	0.9–1.4	25–33	3– 6	bal	30–43	300–400	Brass casting dies and extrusion dies; contact surfaces on internal combustion engine valves
Medium carbon	1.2–1.7	25–33	6–10	bal	40–48	375–650	Hot pressing dies; facing of cutting edges of shears used in chemical, plastics and paper industries
High carbon	2.0–3.0	25–33	10–18	bal	45–60	420–650	Facing of pump sleeves, rotary seals and bearings

Table 11.8 Nickel-boron alloys

Type	Composition, %						Hardness	
	C	Cr	Fe	Si	B	Ni	HRC	HV
Lower carbon	0.3–0.6	8–16	4–5	4–5	2–3	Bal	35–45	330–420
Higher carbon	0.5–1.0	12–18	4–5	4–5	2.5–4.5	Bal	45–62	420–680

Table 11.9 Nickel-molybdenum-chromium alloys

Composition, %						Hardness*	
C	Cr	Mo	Fe	W	Ni	HRC	HV
0.1–0.5	15–18	16–18	5–6	4–5	Bal	24–32	255–310

*Gas welded, in carburising flame

Table 11.10 High molybdenum alloys

Designation	Composition, %					Hardness	
	Cr	Mo	Si	Ni	Co	HRC	HV
T-400	8	28	2	—	Bal		
T-700	15	32	3	Bal	—	52–60	510–650
T-800	17	28	3	—	Bal		

allic (Laves) phase dispersed in a softer matrix. (The hardness of the intermetallic phase is around 1100 HV). The high molybdenum content provides excellent corrosion resistance in the presence of mineral acids. These alloys find use in journal bearings, thrust washers, seals and valves. The nickel based T-700 has a much higher wear coefficient than T-800 (also higher than Stellite 12 and Haynes 716)[4]. The corrosion resistance of some typical nickel and cobalt base alloys is shown in Table 11.11.

(vi) Wear plates and tiles

When wear is likely to occur over large and relatively flat surfaces it can be beneficial to use plate that has already been hard surfaced on one side, or one of a range of solid plate materials in wrought, cast or sintered form. These plates or tiles cover a wide range of hardness, wear and corrosion resistance (Table 11.12)[7] and are useful for lining plant that is subjected

Table 11.11 Comparative corrosion data of some nickel and cobalt base hardfacing alloys[4]

Alloy	Media, concentrations and temperatures			
	30% formic acid, 150°F (66°C)	30% acetic acid, boiling	5% sulfuric acid, 150°F (66°C)	65% nitric acid, 150°F (66°C)
Co base alloys				
Stellite alloy No.1	E	G	S	G
Stellite alloy No.6	E	E	E	U
Stellite alloy No.12	E	G	E	E
Stellite alloy No. 21	E	E	E	E
Stellite alloy No.156	G	S	U	G
Stellite alloy No.157	U	S	U	U
Tribaloy alloy T-400	G	E	G	U
Tribaloy alloy T-800	—	E	G	—
Ni base alloys				
Deloro alloy No.60	G	U	U	U
Deloro alloy No.50	S	U	U	U
Deloro alloy No.40	S	U	U	U
Hastelloy alloy C	E	E	E	S
Tribaloy alloy T-700	E	E	E	S
Haynes alloy No.711	G	E	U	U
Haynes alloy No.716	E	E	U	U

Code:
E = less than 5 mpy* (0.13 mm/y)
G = 5 mpy (0.13 mm/y) to 20 mpy (0.51 mm/y)
S = more than 20 mpy (0.51 mm/y) to 50 mpy (1.27 mm/y)
U = more than 50 mpy (1.27 mm/y)
*mpy = thousandths of an inch per year.

to impact or sliding wear eg chutes, pulverising equipment.

Solid metal or clad plates can, if the carbon equivalent of the bulk or backing material is low enough,

Table 11.12 Characteristics of some wear plates and tiles[7] ex British Steel Corporation Corporate Engineering Standard 23 Pt. 3 1980

Bulk or surface material	Hardness HV or Mohs	Characteristics*				
		Resistance to		Toughness	Resistance to	
		Sliding wear	Impact wear		Heat	Corrosion
Carbon steel (bulk material)	160–260HV	4	4	2–3	4	5
Low alloy steel (bulk material)	250–500HV	3	3	2	4	4–5
Austenitic manganese steel (bulk material)	200–600HV	3–4	2–4	1–2	5	4–5
High chromium iron (submerged arc clad plate)	425–800HV	2	2	3–4	2–3	2–3
Cast basalt	8–9 Mohs	2–3	3–4	5	—	1
Quarry tiles	7–8 Mohs	5	5	5	—	2
Sintered alumina (96% Al₂O₃)	9 Mohs	2	4	5	—	1
Fusion cast alumina (50%Al₂O₃ 32%ZrO₂ 15%SiO₂)	9 Mohs	2	4	5	2	1
WC–Co 80/20 94/6	up to 1800HV	1	2–3	3	3–4	3–4

*1 = Excellent 2 = Very Good 3 = Good 4 = Moderate 5 = Poor

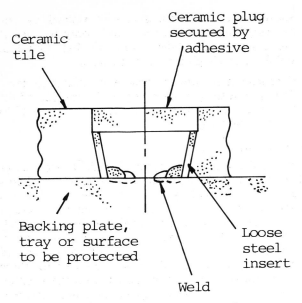

Fig. 11.3   A method for attaching ceramic tiles.[7]

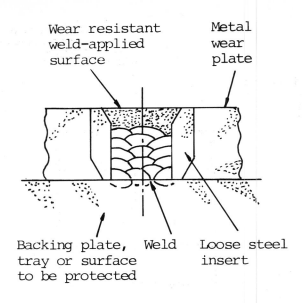

Fig. 11.4   A method for attaching metal wear plate.[7]

be welded directly to the plant. If this is not possible the plates or tiles can be attached by:

—through-bolts with the heads hardfaced by welding;
—hollow metal inserts welded to the plant on their inner surface (Fig. 11.3 and 11.4)[7];
—metal trays in which the tiles are set in cement or adhesively bonded (Fig. 11.5)[7];
—brazing directly to the plant (for tungsten carbide-cobalt tiles).

Where tiles are bonded using cement or adhesive it is beneficial to stagger the joints that lie parallel to the line of motion of the abrading media. Hexagonal tiles are sometimes used to avoid this problem.

It is expensive to cut the harder plates to size and to drill them to receive bolts or inserts. This cost must be taken into account when comparisons are made with other hardfacing systems.

4. Processes and equipment

(i) Introduction

The various hardfacing processes, with their principal characteristics are listed in Table 11.13 and are described in greater detail in the following sections. It is, however, appropriate to consider first four aspects of the technology that have relevance, to a greater or lesser extent, to all processes.

(a) Dilution and contamination effects

The properties of a hardfacing alloy may be modified as a result of compositional changes occurring on fusion with the base metal. The extent of this will depend on the welding process used (see Table 11.5) and on the composition gradients involved. To mini-mise this effect, single layer deposits should be made with low heat inputs, and since the effects are greatly

Table 11.13   Comparison of hardfacing processes

Method	Form of filler	Approx. min deposit thickness, mm	Dilution of deposit, %	Usual mode of application	Deposition rate*, kg/h
Oxy-acetylene, with welding rods	Bare wire, rod or tube	0.5	1 to 5	Manual	½ to 3
Oxy-acetylene with powders	Powder	0.08	1 to 5	Manual	½ to 7
Tungsten-inert gas (TIG)	Bare rod, wire or tube	1	5 to 10	Manual	½ to 2
Plasma transferred arc	Powder	0.25	5 to 30	Fully automatic	½ to 7
Shielded metal arc	Flux coated wire, rod or tube (manual), flux cored wire (semi-automatic)	2	10 to 30	Manual / Semi-automatic	1 to 3 / 2 to 10
Open arc	Tubular wire, which may be flux-cored	2	15 to 25	Semi-automatic	2 to 10
Metal inert gas (MIG)	Bare wire or tube	2	10 to 25	Semi-automatic	2 to 10
Submerged arc	Bare wire, tube or strip	2	15 to 35	Fully automatic	2 to 70
Electroslag	Bare rod or tube	20		Fully automatic	50 to 350

*Overall or effective rate, taking into account any necessary breaks in the deposition process (such as changing electrodes). The gross deposition rate in arc hardfacing is dependent on the welding current.

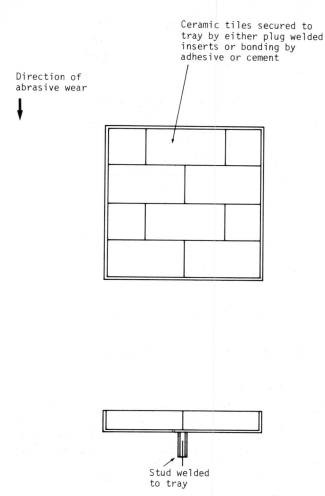

Direction of abrasive wear

Ceramic tiles secured to tray by either plug welded inserts or bonding by adhesive or cement

Stud welded to tray

Fig. 11.5   Method for attaching ceramic tiles in trays.[7]

reduced in the second and subsequent passes, at least two-layer deposits are recommended for most hardfacing operations. Arc welding with dc straight polarity (electrode negative), with its reduced penetration, generally results in less dilution than welding with reverse polarity (electrode positive).

(b) Substrate considerations

The method of application of hardfacing depends to a large extent on the composition of the substrate.

Low and medium carbon steels (<0.4%) can be readily hardfaced by any of the methods in Table 11.13. As the carbon content increases above 0.4% greater attention has to be paid to heating prior to welding to reduce the tendency to form martensite in the heat affected zone. Similar precautions should be taken with martensitic stainless and tool steels which, in addition, need control of interpass temperature and cooling rate.

The harder die, tool and age-hardenable steels need to be annealed or overaged before, and cooled at a controlled rate after, hardfacing.

Austenitic manganese steels should be kept as cool as possible or heated and cooled rapidly through the temperature range 400–600°C, for this reason arc welding is preferred.

Austenitic stainless steels can be welded by any process, but oxy-acetylene should only be used if the austenite is stabilised with niobium (titanium stabilised steels give porous coatings).

Cast iron substrates are generally difficult to hardface, pre-heating being necessary if cracking of the hardface is to be avoided. Fluxes are often used to promote bonding.

(c) Crack avoidance

Some of the less ductile hardfacing alloys are susceptible to cracking, aggravated by thermal gradients present during welding or by differences in thermal expansion between the base metal and the overlay. The cracks may have their origin in the base metal, particularly if this is one of the less weldable steels, in which case the usual precautions of preheating and slow cooling should be observed. The need for these will depend on the mass of the part, on the location and amount of hardfacing applied, and on the hardenability of the base metal. The wear-resistant properties of a great many hardfacing alloys depends however on their martensitic structure, which in certain cases is only developed on rapid cooling from the welding temperature. Consequently the use of low alloy martensitic steel fillers should be restricted to the hardfacing of the more readily weldable steels.

The harder surfacing alloys will tend to crack if attempts are made to deposit them in thick layers. Where therefore a heavy build-up is necessary, it is usual to start by depositing 'buttering' layers of a cheap but tough material, covering these finally with one or two layers of the wear resistant alloy. Apart from reducing the tendency to cracking, this procedure is clearly desirable in that it conserves the more expensive material. Prior to the deposition of the buttering layers, it is sometimes an advantage to cut parallel grooves across the surface to be treated, using a grooving electrode. This increases the bonding area and provides better stress distribution.

(d) Safety

The safety precautions adopted in hardfacing are those appropriate to the particular welding process in use. Owing to the relatively low voltages employed in electric arc welding, the danger due to electric shock is minimal. Ultra violet radiation hazards are more significant, particularly with open arc and tungsten inert gas methods, when protection of the skin as well as the eyes is important. With metal arc and metal inert gas methods radiation is less but eye protection is still necessary. Gloves should be worn where possible, especially if the process produces spatter. The radiation hazard is absent in the case of submerged arc welding.

(ii) Oxy-acetylene process

This technique involves heating the substrate in an oxy-acetylene gas flame and then melting the coating on to it using a metallic filler rod, with a flux to secure a bond. Alternatively, when surfacing a ferrous base with one of the highly oxidation resistant cobalt-base alloys, the flux may be dispensed with and the

151

base heated using a flame containing an excess of acetylene; the carburising conditions then cause the formation of a very thin high carbon low melting point layer on the surface of the steel. This layer, visible as 'sweating', forms an excellent metallurgical bond with the hardfacing alloy when this is melted into it. The deposits generally pick-up less than 5% iron from the steel base so that there is virtually no deterioration in the properties of the hardfacing alloy due to dilution effects.

The technique employs low cost, widely available equipment and can be used for a broad range of applications with various combinations of base metal and surfacing alloy. The equipment is readily portable and requires no electrical power supply. It is mainly used however with the high carbon, high chromium filler rods and with the more oxidation resistant cobalt- and nickel-base alloys. In skilled hands it can produce surface coatings of very high quality and is used wherever good depositional control is important, such as on cutting edges and thin section components. Its principal limitations are its low deposition rate and, for the best results, the need for a high degree of operator skill.

Where the nature and quality of work warrants it, mechanised welding heads may be used, the filler wire being automatically fed to the fusion zone by means of guide rolls. This, and the use of multiflame burners, can considerably increase the speed of the process.

A variation of the oxy-acetylene process employs powders in place of filler rods, the powder being fed to the torch by aspiration into the fuel gas from a suitable feed control device. Known as the powder welding method, this has the advantage of being able to use alloys that are not readily available in rod or wire form, (but which can be produced as powders). Using nickel- or cobalt-base self-fluxing alloys, it is possible by this means to produce thin, porosity-free deposits, even with relatively unskilled operators, and the lower temperature of deposition tends to give less scaling and distortion than is liable to occur using solid filler rods.

Whereas the powder welding process is admirably suited to the precision hardfacing of small parts, it is too slow for application to large areas. These are better dealt with using the two-stage spray-fuse process, in which the powders are first sprayed on to the part using a conventional metal spraying gun; the relatively unconsolidated coatings are then fused to the substrate using an oxy-acetylene torch, a controlled atmosphere furnace or an induction heating coil. (Both the powder welding and spray-fuse methods are more fully described in Chapter 10).

(iii) Tungsten inert gas (TIG) process

In this process the filler rod or wire is fed, usually manually, into an arc struck between a non-consumable tungsten electrode and the workpiece, the molten pool being protected from oxidation by a flow of inert gas, generally argon. (Fig. 11.6). The equipment needed is more expensive, and less portable than for oxy-acetylene, requires a similar degree of skill but, owing to the protective gas shield and

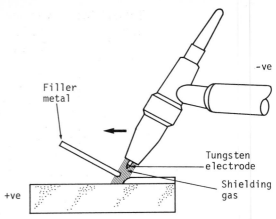

Fig. 11.6   Tungsten-inert gas (TIG) process.

the higher temperatures attainable, allows a wider selection of surfacing alloys. The need to preserve the gas shield intact means that the process has to be carried out in-doors. The process tends to be used mainly for surfacing operations involving the high alloy nickel- and cobalt-base alloys, and for producing high quality deposits on small areas such as in grooves or recesses. It is especially useful in situations where a high heat input is of advantage, and for surfacing austenitic stainless steel with cobalt-base alloys, when the carburising conditions normally employed in gas welding tend to give rise to grain boundary films of chromium carbide and loss of corrosion resistance. The deposits are, however, more subject to base metal dilution than those made by the oxy-acetylene method, and care is necessary if over-softening due to this effect is to be avoided. Little post deposition machining is necessary, an advantage when depositing very hard materials.

The slow rate of deposition associated with manual operation can be improved if the process is automated when, by controlling the wire feed, width of oscillation and speed of travel, a very uniform deposit is obtainable. The use of dc straight polarity is normal practice.

(iv) Plasma transferred arc (PTA) process

This process uses a direct current, tungsten inert-gas arc within an arc-restricting orifice, the plasma forming as the gas flowing through the orifice ionises. The hardfacing metal in the form of powder is fed to the ionised gas stream (Fig. 11.7)[8] as it leaves the restriction; supplementary shielding gas can be added as a protective shroud. A second dc supply is connected between the electrode and the arc-restricting orifice supporting a non-transferred arc that provides additional energy and serves as a pilot arc to initiate the transferred arc. Deposition rates of up to 6 kg/h are obtainable. The process is ideally suited for use in automated systems.

The process is mainly of interest for parts which have to be precision surfaced with relatively brittle materials, such as the nickel-chromium or cobalt-chromium alloys. A wide range of these are now available in powder form, often containing boron and silicon additions to provide self-fluxing characteristics. If desired, tungsten carbide particles may be mixed

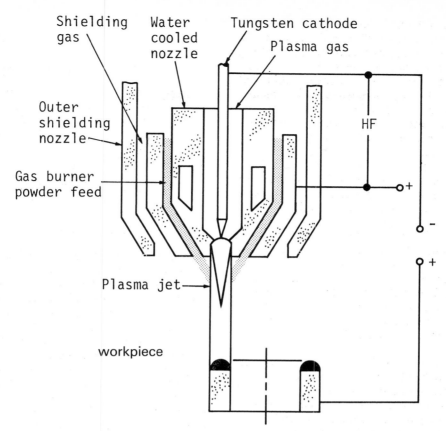

Fig. 11.7 Plasma transferred arc (PTA) process.[8]

with the alloy powder to form heterogeneous deposits having exceptional abrasion resistance; if however, deposits exceeding about 1.6 mm thickness are required, it is usual to arrange for the carbides to be separately fed to the weld pool directly behind the arc. This avoids the danger of the heavy particles settling to the bottom of the weld pool and minimises the extent to which carbide melting and consequent matrix embrittlement occurs.

The process is capable of producing very flat, smooth deposits, requiring very little if any finish grinding. It is especially suited to the surfacing of regularly shaped components where the volume of work justifies the installation of fully automated equipment costing in the region of £30,000–40,000. In practice, the operator controls shielding gas flow, powder gas flow, torch orifice, work distance and current settings; preheat, weld start and finish, and torch movement may however be automatically programmed.

A number of companies manufacturing high duty control valves for the process and nuclear industries now employ this process in place of oxy-acetylene or TIG welding, mainly for the Stellite facing of stainless steel valve plugs and seat rings.

(v) Shielded metal arc process (sometimes referred to as manual arc or stick welding)

In this process the electric arc is struck between a consumable flux-covered electrode and the workpiece, the molten metal being protected from atmospheric contamination by the gaseous shield evolved during combustion and decomposition of the electrode coating and by molten flux formed over the weld pool, (Fig. 11.8).[2] Dilution of the deposit by the base metal occurs but this can be minimised by operator skill, by restriction of the welding current and by the use of dc straight polarity (electrode negative); dilution of the first run is however never less than about 10%. The high rate of heat input compared with the oxy-acetylene process permits higher deposition rates and the hardfacing of comparatively heavy sections that would require an excessive time to heat up using a gas torch. The equipment required is readily available and inexpensive.

A wide range of iron-base alloy electrodes is available for hardfacing purposes, the alloying constituents being contained either in the core of the electrode, in its flux covering or, less usually, as a powder in a tubular electrode. (The latter method is used where the alloy is difficult to fabricate as a solid rod.) The flux coatings are generally based on lime-fluorspar and, when properly dried before use, yield deposits having a low hydrogen content and a reduced tendency to cracking.

The process finds its greatest application where coatings can be used as-deposited, ie without further machining or grinding. Because of the dilution that occurs, the process is mainly used for hardfacing with the cheaper alloys. It is however an extremely versatile and flexible process, eminently suited to jobbing work and to maintenance and field work. Like other manual methods, however, the deposition rate is low, and where large areas have to be covered, the use of one of the semi-automatic methods is to be preferred.

153

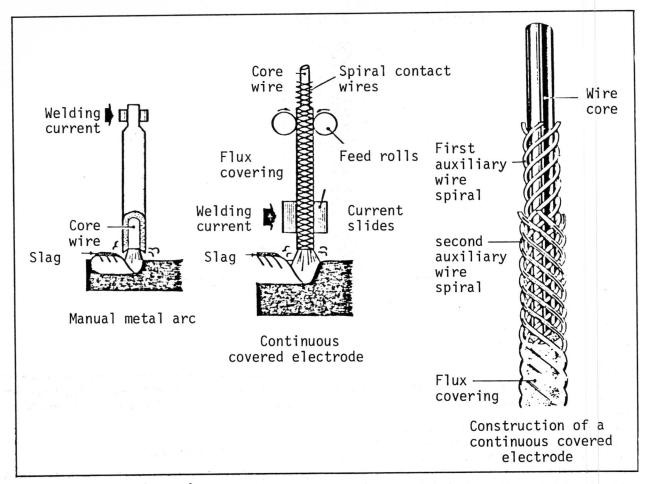

Core wire Spiral contact wires

Welding current

Flux covering Feed rolls

Welding current Current slides

Core wire

Slag Slag

Manual metal arc

Continuous covered electrode

Wire core

First auxiliary wire spiral

second auxiliary wire spiral

Flux covering

Construction of a continuous covered electrode

Fig. 11.8   Shielded metal arc process.[2]

Electric power consumption is between 1.5 and 2 kWh/kg of metal deposited, figures which also apply to most other consumable electrode hardfacing operations.

(vi) Open arc process

This is a semi-automatic process in which the electrode, in the form of continuously coiled tube, is automatically fed to the arc, its movement being controlled by electrical feed-back from the arc voltage. The operator has only therefore to control the movement of the arc and the deposition of metal, so that less demands are placed on his skill then when using the shielded metal arc process. The core of the tube contains the alloying elements together with deoxidising material, the latter being necessary because the process employs no shielding gas or flux protection. A more recent development, however, which has considerably extended the application of this process, is the flux-cored wire, in which a measure of protection from oxidation is provided by gas-forming minerals in the core of the tubular electrode (Fig. 11.9).[2]

The method is very widely used for hardfacing and, of the semi-automatic methods, requires the least outlay on equipment, only a standard power source and wire feed unit. Higher welding currents can be used than with the shielded metal arc process, allow-

Fig. 11.9   Open arc process—sections through flux cored wires.[2]

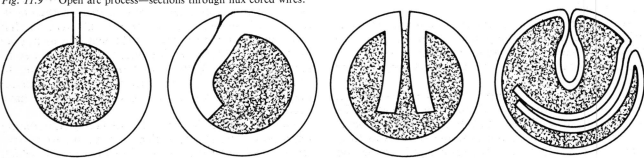

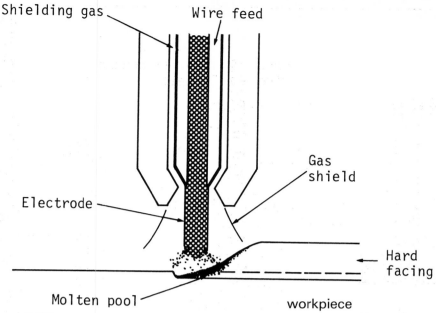

Fig. 11.10 Metal-inert gas (MIG) process.

ing higher deposition rates which in practice are further increased by the improved productivity (ie arcing time) that the automated electrode feed system permits.

The method produces more spatter than other arc welding processes, and the deposits are not of such high quality. They are nevertheless perfectly adequate for many hardfacing operations and, with the introduction of flux-cored electrodes, the process is gaining in popularity, especially for site work. It lends itself to full automation where the cost of this is warranted.

(vii) Metal inert gas (MIG) process

This process is essentially open arc welding with the added provision of inert gas shielding, to prevent oxidation and alloy losses (Fig. 11.10). The shielding gas is usually either argon or carbon dioxide, with a small quantity of oxygen added to improve arc action and metal transfer. (When using shielding gases containing oxygen, welding wires containing a deoxidant must be employed). The use of shielding gas adds to the cost of the process, and MIG surfacing should in general only be used in preference to the open arc process where the extra cost is warranted by the need for improved deposit quality. In certain cases the use of CO_2 shielding modifies metal transfer characteristics in such a way that, on these grounds alone, its use is to be preferred.

Whereas most MIG hardfacing operations are done using semi-automatic equipment, the method may be fully automated for surfacing large areas of regular shapes; deposition rates of up to 25 kg/h or more may then be obtained.

A variation of the process involves feeding an extra filler wire to the weld pool, so increasing the deposition rate while at the same time reducing penetration and dilution by the base metal. This method may have advantages over the submerged arc method

in situations where high deposition rates are required but where suitable granular flux materials are not available for the particular alloy it is desired to deposit.

(viii) Submerged arc process

In this process the protective gas is replaced by a granular flux which surrounds the arc and covers the molten pool (Fig. 11.11a). In addition to providing protection from oxidation, the flux, after solidification, thermally insulates the coating and so allows it to cool more slowly. The alloy content of the deposit is controlled by the composition of the filler rod and by metallic additions made to the flux. The filler material may be in the form of wires, rods, tubes or strip. The need to keep the powdered flux in place means that the process must be operated in the down-hand mode onto flat or rotated cylindrical surfaces.

The process is invariably fully automated and, by using higher currents than are possible with hand-held equipment, high deposition rates are attainable. It is therefore applicable to the surfacing of large areas, such as steel mill roll faces, when a consistently high quality deposit can be obtained without the need for a high degree of operator skill. The deep penetration of the arc, together with the insulating effect of the flux blanket, is responsible for the development of intense heat in the welded area, so that some preheating and/or post weld heat treatment may be necessary. Moreover the intense heat gives rise to marked dilution of the deposit by the base metal, so that the full protective properties of the filler material may not be realised until two or more layers have been deposited.

The use of strip electrodes for surfacing is becoming important. The strips, which may be up to 80 mm wide, may be of rolled or powdered material in which

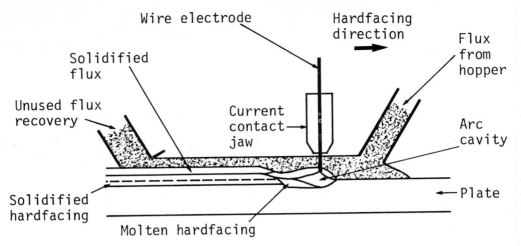

Fig. 11.11 Submerged arc process.

re-inforcing wires are embedded. During deposition the arc oscillates rapidly across the width of the strip, which then assumes the contour of the workpiece, so allowing the deposition of a uniformly thick layer of hardfacing material. Deposition rates can be very high, up to 40–50 kg/h. For a given current, the use of dc straight polarity (electrode negative) produces the highest deposition rates with the least penetration.

A variation of the process, developed by Texas Alloy Products in the USA and operated under licence in the UK by Trimay Engineering Co. Limited, is referred to as the Bulkweld process. In this, as shown schematically in Fig. 11.12, powdered filler alloy is introduced between the workpiece and the electrode, the arc melting both the work surface and the powder. The required thicknesses are generally built up by depositing successive layers, each approximately 4 mm thick.

Owing to the buffering action of the powder, penetration is very shallow and relatively little dilution of the alloy deposit occurs, 2% to 5% in the first layer, as compared with 15% to 35% with conventional submerged arc. The process is capable of applying deposits at rates comparable with those obtainable using the conventional process, and although originally used for repair and reclamation, is now exploi-

Fig. 11.12 Submerged arc process (bulkweld version).

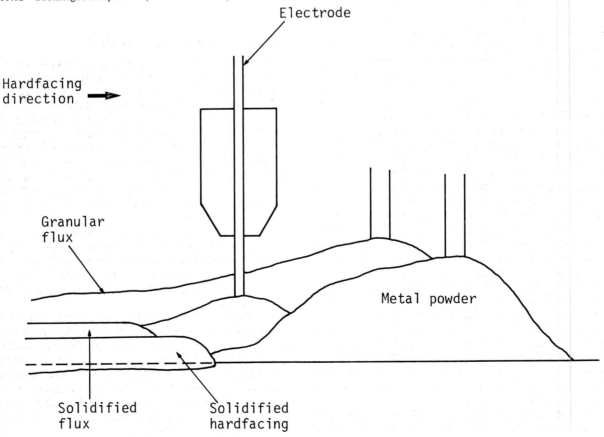

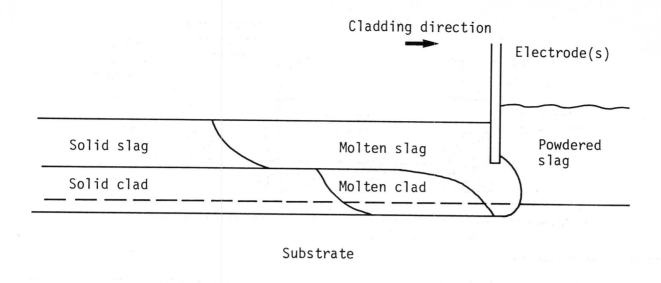

Fig. 11.13 Electroslag process.

ted in the manufacture of original components, mainly for steelworks plant, (eg work rolls, table rollers, edge and pinch rolls, crane wheels, shafts and components for ore and sinter plants). It is also used in the manufacture of pre-coated wear plates, usually for the application of chromium carbide rich steels.

(ix) Electroslag process

This process differs from the submerged arc process in that the electrode and substrate are melted, not by an electric arc, but by immersion in slag maintained in the molten condition by the passage of an electric current through it. The molten bath, which can be maintained on a vertical or horizontal surface, is retained by water cooled copper moulds or sliding shoes which can be moved over or along the workpiece as solidification of the metallic deposit proceeds. Single or multiple electrodes are used and are generally given a reciprocating motion; alternatively if a single electrode of heavy cross section is employed, electrode oscillation can be omitted (Fig. 11.13).

Under properly selected conditions the process exhibits a high degree of stability, even at low current densities. Arc formation is discouraged by the use of low open circuit voltages and a slag of high electrical conductivity (usually fluorspar based). Deposition rates can be very high, around 10 kg/h per electrode. Thus, using 10 electrodes, with a total current consumption of 5000–6000 A, the deposition rate may be up to 100 kg/h. Surfacing by these means is a single stage process in which deposits up to 60–70 mm in thickness may readily be obtained. Thick deposits of high alloy surfacing materials may however suffer from dilution effects, a factor which may restrict the deposit thickness. The process is less flexible if strip electrodes are used because of the difficulty in obtaining strip in some compositions.

Electroslag surfacing has potential for development as a means of cladding or re-surfacing steel mill rolls,

for which purpose the submerged arc welding process is used almost exclusively. The process has been developed in Russia and Czechoslovakia, mainly for rebuilding plain barrelled and Pilger rolls. Kawasaki Steel in Japan has recently developed the Maglay process, in which the flow of melt is controlled by the imposition of magnetic fields (Fig. 11.14).[9] The process could be modified to deposit hardfacing materials on large surface areas such as rolls.

(x) Paste fusion process (boriding)

A number of proprietary 'sweat-on' pastes are available which, when applied to a surface with a suitable spreader and allowed to air dry, can be fused to the surface using an oxy-acetylene, TIG or carbon arc torch. Fusion can be carried out at relatively high speeds to produce very smooth coatings of high hardness, typically 60–65 HRC. The paste is normally applied in a layer approximately 1.5 mm thick which, on fusion, results in surface hardening to a depth of 3 to 4 mm. The fusion stages can with advantage be automated, when a more uniform depth of penetration is obtained.

The pastes, which are usually water-based contain a high proportion of chromium borides, together possibly with tungsten carbide, in a self-fluxing nickel-base matrix. Applied to a steel surface, they yield coatings consisting of hard borides dispersed in an alloy steel matrix.

Coatings of this type are particularly resistant to fine particle erosion, and have been used with success in pulverised fuel lines and for facing blades on quarry and cement kiln fans. The life of exhauster fan blades on a fuel pulveriser treated using this process are reported to have been doubled, as compared with unprotected mild steel blades. (Brazed on tungsten carbide wear tiles are better, but are considerably more expensive).

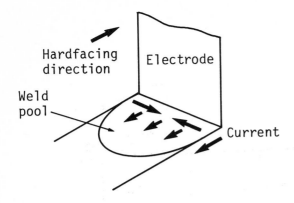

(a) In strip submerged arc and electro-slag welding under-cutting is caused by a flow of material towards the centre of the weld pool under the influence of forces developed by the welding current.

Fig. 11.14 The Maglay electroslag process.[9]

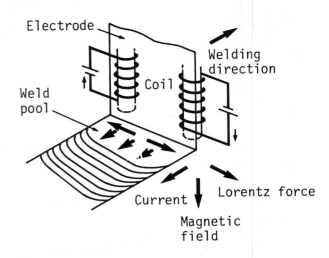

(b) In the Maglay process material flow is controlled by an applied magnetic field.

(xi) Brazing

Although strictly not a welding process, brazing can be used to fix coatings onto metal substrates either by the attachment of, for example, wear plates made of tungsten carbide-cobalt or by attaching tungsten carbide particles directly to the substrate. These processes can be carried out using a localised heat source such as oxy-acetylene or an arc process, or by heat treatment in a furnace. The former has less effect on the mechanical strength of the substrate, the latter is more expensive but can give more uniform results.

Brazing can also be used to attach metal cloths or tapes containing hard wear resistant particles eg Conforma Clad produced by Imperial Clevite Inc.[10]. The method can be used to give coatings up to about 2½ mm thick on complex shapes that will resist low stress abrasive wear and metal-to-metal contact.

5. Costs

(i) Equipment

The cost of oxy-acetylene equipment is low: a torch and gas leads costing £350–£400. Electric arc equipment for manual welding operations consists of a welding generator, power cables and electrode holder: total cost is possibly about £1500. Equipment for semi-automatic or MIG welding is more expensive: approximately £3000 including power source and wire feeding unit.

Fully automatic equipment tends to be purpose built, its cost being determined by the type of surfacing operation it is required to perform. For fully automatic surfacing of, say, valves or valve seats, equipment costs in the region of £25,000. Equipment for the automatic surfacing of steel mill rolls weighing up to 35 tonnes may cost in the region of £60,000.

This expense is clearly only justified where hardfacing is being used for routine restoration or manufacturing operations.

(ii) Consumables

The cost of consumables varies widely, depending on the type, composition and gauge, as well as on the quantities in which they are purchased. The costs given in Table 11.14 provide a rough guide to the price of consumables of the compositional types listed in Table 11.1 when purchased in bulk.

When using inert gas welding methods the cost of gas must be taken into account, the current prices of argon is about £2.4/m³. Since MIG welding uses 0.25–0.5m³/kg of metal deposited, the cost of gas is in the region of £0.6–£1.12/kg of metal deposited.

(iii) Hardfacing

The cost of hardfacing, in terms of metal deposition rate, is given in Fig. 11.15. This diagram assumes that:

(a) consumable costs are as indicated in Table 11.14;
(b) a metal recovery rate of 90% is achieved;
(c) the labour plus overhead rate is £15/h; and
(d) the equipment is in use for 70% of the time.

Table 11.14 Approximate cost of consumables

Compositional type	Cost £/kg
Austenitic manganese steels, martensitic and high speed steels, martensitic and high chromium irons	2–16
Nickel base alloys	15–35
Cobalt base alloys	25–50
Tungsten carbide + iron matrix	30–40
Tungsten carbide + Ni/Co matrix	40–60
Titanium carbide + iron or cobalt alloy matrix	55–70

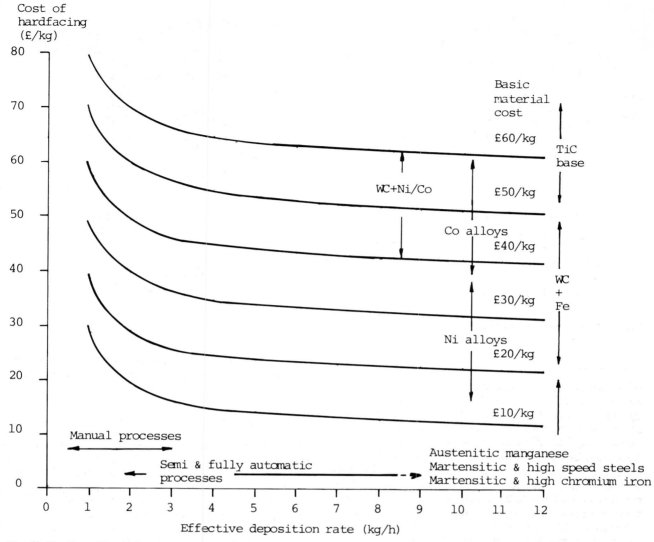

Fig. 11.15 Cost of hardfacing.

It is clear that, for effective deposition rates in excess of about 3 kg/h, the direct cost of hardfacing is dominated by the cost of consumables. The cost of hardfacing is easy to compute if only the labour costs for the time taken to deposit the coating are taken into account. (Table 11.15). These, however, are never the total cost of hardfacing. If the case of hardfacing a simple rod 50 mm dia. × 200 mm long is considered it will be necessary, to ensure that the rod is square ended after treatment, to start with a rod of greater diameter and length and to machine a depression over a length of about 210 mm into which

the hardfacing can be deposited. The rod would probably have to be pre-heated before hardfacing starts and, to avoid overheating, a rate of deposition less than the capability of the deposition process may have to be used. Most welding processes leave a relatively rough surface which must be machined (or ground if the hardness is above 58 HRC ie 680 HV) and these costs together with the cost of the extra material deposited as a grinding allowance must be taken into account.

Depreciation of equipment is probably not a significant proportion of the total cost of hardfacing unless

Table 11.15 Marginal costs for labour and materials for hardfacing a 50 mm dia × 200 mm long bar*

Hardfacing material	Material cost (£/kg)	Process	Effective deposition rate (kg/h)	As deposited coating thickness (mm)	Material cost (£)	Labour cost (£)
Iron base alloy	3	Manual TIG	1	4	7½	53½
Nickel base alloy	30	semi automatic MIG	4	4	94	17
Cobalt base alloy	40	plasma transferred arc	5	4†	113	12
WC–Iron base	40	oxyacetylene	2½	4	125	26½

*Excluding setting up, pre-machining, pre-heating, grinding and equipment depreciation. It is assumed that labour costs are £15/hour, equipment is used for 70% of the time and the deposit efficiency is 100%.
†The plasma transferred arc gives a deposit with a smooth surface that requires less surface finishing.

equipment such as plasma transferred arc, complex submerged arc equipment, or expensive manipulative equipment is involved.

Several published articles [5,11] cover the economics of hardfacing.

6. Current developments

Because hardfacing by welding is intimately concerned with the economics of manufacture and reclamation, the development of more cost effective processes continues to attract commercial interest. There is an incentive to develop new hardfacing alloys and to provide means of depositing wear resistant surfaces at faster rates, at less cost, with less dilution and having a surface quality that reduces the cost of final machining or grinding. There is also a trend to develop processes that are more amenable to automatic and programmed operation, with the use of robots.

Hardfacing using a laser beam as the heat source is being evaluated with the objective of improving deposit quality by controlled dilution from the substrate. Thin deposits, which need little post-weld machining or grinding, can be produced by this means. Both wire and powder feed have been explored and result in better metallurgical control and reproducibility than TIG or MIG. Narrow Stellite and Tribaloy 100 tracks have been successfully deposited on Nimonic 75.[12] Claims by Russian workers that Stellite can be deposited by friction cladding[13] has prompted more recent work into this process. It gives good bonding with apparently no dilution.

On the material side there is a need to improve hardness and wear resistance while using less scarce and strategic materials. Thus there is a move to develop tungsten-free cemented carbides which have similar fracture toughness and abrasive wear resistance to conventional tungsten carbide-cobalt alloys[14] and to develop cermets based on titanium carbide and vanadium carbide rather than tungsten carbide. The relatively low rupture strength of titanium carbide based hard metals has been shown to be improved by additions to the titanium carbide (in solid solution) of carbides of zirconium, hafnium, vanadium, tantalum, molybdenum and tungsten.[15] Attempts are being made to improve the hardness of carbides and of the matrix in the 3.5/4.5% carbon, 25/30% chromium irons by adding boron.[16]

References

1 DAVIES A C. *The Science and Practice of Welding*. Cambridge University Press London, 6th edn. 1981.

2 HOULDCROFT, P T. *Welding Process Technology*. Cambridge University Press London 1979.

3 Metals Handbook (ASM) Vol. 6, *Welding, Brazing and Soldering*, 1983.

4 Deloro Stellite—trade literature.

5 GREGORY, E N. The economics of surfacing *Surfacing Journal,* Vol. 14, (1), p 2, 1983.

6 Metallurgical Industries Inc.—trade literature.

7 British Steel Corporation Corporate Engineering Standard 23 Pt 3, p 5, 1980.

8 Messer Griesheim Trade Literature.

9 NAKANO, S. *et al.*, Maglay—an electroslag technique for overlay welding. *Surfacing Journal*, Vol. 14, (2), p 23, 1983.

10 SHEWELL, D E. New method of applying wear resistant coatings, *Metal Progress*, p 41, Nov. 1983.

11 HAGUE, F. *et al.* The effectiveness of hardfacing—the analysis of fields trials. *Australian Welding Journal,* p 16, March/April 1978.

12 COURTNEY, C and STEEN, W M. Surface coating using a 2 kW CO_2 laser. *Laser 1979 Conference Proceedings,* Munich, p 228.

13 NOGK, M V. *et al.* Friction Hardfacing of Steel with Stellite V3K, *Svar Prioz,* No. 7, p 16, 1970.

14 HOLLECK, H, PRAKASH, L AND THUMMLER F. Molybdenum based, tungsten free Cemented Carbides, *Metal Powder Report,* Vol. 36, (12), p 583, 1981.

15 HOLLECK, H. The Constitution as the Basis for Recent and Future Developments of Wear Resistant Material. *Specialty Steels and Hard Materials*, p 383–388, Pergamon Press 1983.

16 BARNS, H. Wrought Cast and Weld Deposited Abrasion Resistant Alloys with Superhard Phases, *International Conference on Tribology in Mineral Extraction,—War on Wear*. Nottingham p 193, 1984.

Additional reading

Welding '84—This buyers guide published annually by The Welding Institute lists equipment and consumable suppliers, also 'subcontract' organisations.

Proceedings of seminars held by The Welding Institute
1. *Weld surfacing and hardfacing.* 1980
2. *Selecting the right surfacing alloy for the job.* 1983.

Surfacing Journal—This quarterly publication by The Welding Institute contains topical articles plus abstracts on surfacing papers and patents.

APPENDIX 1

Wear Test Methods

There exists a large variety of wear testing methods; ASLE[1] lists some 270 types of friction and wear devices. Only a more limited number of test methods are used commonly in numerous laboratories. Details of the wear test methods which are referred to in this publication are given on page 1. Further details of the test methods can be obtained from the relevant references.

Although wear data is quoted in various sections of this publication it must be emphasised that the order of merit of different coatings or treatments obtained in a particular type of test may only be applicable to a limited range of service conditions. In consideration of a service application using laboratory data great care should be taken in assessing whether a particular test method effectively simulates the application. Even in cases in which laboratory simulation is representative prototype testing is recommended prior to the adoption of a treatment or coating in production.

References

1 Friction and Wear Devices, ASLE.

2 SUDDABY, O, *Tribology News,* Inst. Mech. Engrs. 26, 1975.

3 Amsler machine—Manufactured by Alfred J Amsler & Co. Schaffhausen, Switzerland.

4 SALTZMAN, G A, Plastics Design and Proc. October 1975.

5 KEDWARD, E C, *et al. Tribology International,* October 1974.

6 LONGO, F N & FORANT, H, *AWS 8th Thermal Spraying Conf.* 1976.

7 PARKER, K. *Plating* 61, pp 809–892, September 1974.

8 CHESTERS, W T, *Inst. Mech. Engrs. Conference on Gearing* 1958.

9 SHEEHAN, J P & HOWES, M A H, *SAE Trans.* 81 pp 1031–1045, 1972.

10 Institute of Petroleum Test Method IP 166.

11 FRG Standard DIN 51354.

WEAR TEST METHODS

	FALEX (2)	AMSLER (3)	LFW-1 (4)	PIN AND DISC	HAMMER WEAR (5,6)
TEST CONFIGURATION					
TEST PURPOSE	Determination of the wear rate, load carrying capacity and friction coefficient for sliding contacts.	Determination of the sliding wear rate of test materials, treatments or coatings.	Determination of the adhesive wear rate of materials against an SAE 4620 steel ring.	Determination of sliding wear rate	Determination of cohesion and adhesion qualities of coatings and also their wear characteristics under arduous vibrating conditions.
SPECIMENS	A 6.35 mm diameter cyl-indrical journal is rotated within two loaded stationary V-blocks to give a 4-line contact.	Normally two discs 40 mm diameter and 10 mm thick. Alternatively the discs may be made 30 and 50 mm diameter.	The steel ring is rotated with a stationary block (which may be a coating) in loaded contact.	A pin rubbing on a rotating circular plate.	Obliquely cut specimens of 23 mm² contact area.
TEST CONDITIONS	Journal speed 290 rev/min Sliding velocity 0.1 m/s Load (constant or increasing) 89-20,000 N Hertz stress 242-3450 MN/m² Specimens can be immersed in oil or other fluid.	Lower disc speed 400 rev/min Upper disc speed 440 rev/min Load 196 N. May be run dry or lubricated	Ring speed (max) 197 rev/min. Load (typical) =340 N Load (max) 2800 N Duration 20,000 cycles.	Specific to type of test.	Combined impact and sliding Sliding distance: 0.76 mm Impact frequency: 60-70 Hz Impact load: 60-223 N Temperature: 600°C Duration: 6-10 hr
MEASUREMENTS	Journal driving torque Progressing wear depth Wear life. Seizure failure load Final specimen weights	Individual disc weight loss. Total disc weight loss.	Test block weight loss. Scar width. Ring weight loss.	Weight loss of disc. Height loss of pin.	Volume loss Fretting resistance =1/volume loss
WEAR TYPES	Mild and severe adhesive wear Scuffing	Mild and severe adhesive wear. Machining wear.	Mild and severe adhesive wear.	Mild and severe adhesive wear	Adhesive wear Fatigue Fretting

162

WEAR TEST METHODS

	DRY RUBBING WEAR (5)	TABER ABRASION (7)	WET SLURRY ABRASION (6)	ROLLER CONTACT FATIGUE (8, 9)	GEAR SCUFFING, IAE (FZG) (10, 11)
TEST CONFIGURATION					
TEST PURPOSE	To evaluate over a broad temperature range, materials used in gas turbines.	Determination of the abrasive wear rate and the Taber wear index of material surfaces.	Determination of the abrasive wear rate and the abrasion resistance of materials and treatments.	Determination of the rolling contact fatigue life of materials and treatments used for gears.	Determination of the load to cause scuffing of test gear material/lubricant combinations.
SPECIMENS	Spherical buttons against a flat pad.	Flat face of the disc abraded by two CS-10 rubber bonded abrasive wheels.	Two specimen pads loaded against a rotating circular plate.	Roller of 102 mm diameter in the test material loaded against and driven by a 25 or 51 mm diameter carburised shaft.	Gears made from test material.
TEST CONDITIONS	Sliding amplitude: 0.254 mm Sliding frequency: 25 Hz Load: 5.57 N Temperature: 20-800°C Duration 10^5 cycles.	Load 9.81 N Wheels cleaned with abrasive paper every 1000 revolutions Test is run without lubricant	Load per pad 112 N Runs in bath of abrasive slurry	Speed 1050 to 1500 rev/min Load variable -Lubrication by spray fed gear oil	Gear tooth 15/16 (16/24) Pinion speed 2000 (2170) rev/min Load increased in equal steps (log. steps) Temperature 60°C (90°C) Lubrication by oil spray (oil bath)
MEASUREMENTS	Volume loss.	Weight loss Taber Wear Index = weight loss/1000 revs (mg)	Volume loss.	Cycles to first fatigue pit. Endurance limit = stress to produce pitting in 10^7 cycles.	Load to produce scuffing.
WEAR TYPES	Adhesive wear. Fretting.	Low stress abrasion.	High stress grinding abrasion Low stress abrasion Machining wear	Contact fatigue (pitting).	Scuffing.

163

APPENDIX 2

Material Specifications

1 Steel specifications British standards

En	BS 970	Composition, main elements
3	070–M20	0.16–0.24C, 0.5–0.9Mn
3A	070–M20	0.15–0.25C, 0.4–0.9Mn
3B	070–M20	0.25C max, 1.0Mn max
8	080–A40	0.36–0.44C, 0.6–1.0Mn
8D	080–A42	0.4–0.45C, 0.7–0.9Mn
9	—	0.5–0.6C, 0.5–0.6Mn
10	—	0.5–0.6C, 0.5–0.8Mn, 0.5–0.8Ni
11	526–M60	0.55–0.65C, 0.5–0.8Mn, 0.5–0.8Cr
12	503–M40	0.36–0.44C, 0.7–1.00Ni
15B	120–M36	0.32–40C, 1.0–1.40Mn
16C	605–A37	0.32–0.4C, 1.3–1.7Mn, 0.22–0.3Mo
18	530–M40	0.36–0.44C, 0.9–1.2Cr
18A	530–M40	0.27–32C, .65–.8Mn, .85–1.15Cr
19	709–M40	0.36–0.44C, 0.9–1.2Cr, 0.25–0.35Mo
24	817–M40	0.36–0.44C, 1.0–1.4Cr, 0.25–0.35Mo, 1.3–1.7Ni
32	045–M10	0.07–0.13C, 0.1–0.4Si, 0.3–0.6Mn
34	665–M7	0.14–0.2C, 0.2–.3Mo, 1.5–2.Ni
36A	655–M13	0.1–.16C, .7–1Cr, 3–3.75Ni
36C	832–M13	0.1–.12C, 0.6–1.1Cr, .1–.25Mo, 3–3.75Ni
39A	659–M15	.12–.18C, 1.–1.4Cr, 3.9–4.3Ni
40B	722–M24	0.2–0.28C, 3–3.5Cr, 0.45–0.65Mo
40C	897–M39	0.35–0.43C, 3–3.5Cr, .8–1.1Mo, 0.15–0.25V
41A	905–M31	0.27–0.35C, 1.4–1.8Cr, .15–.25Mo, .9–1.3Al
41B	905–M39	0.35–0.43C, 1.4–1.8Cr, .15–.25Mo, 0.9–1.3Al
43B	080–A47	0.45–0.5C, 0.7–0.9Mn
52	—	.45C, 3–3.75Si, 0.3–0.6Mn, 0.6 max Ni, 7.5–9.5Cr
59	—	.78C, 1.75/2.25Si, 0.2–0.6Mn, 1.15–1.65Ni, 19/20.5Cr
100	945–M38	0.35–.45C, 1.2–1.5Mn, 0.5–.1Ni, 0.3–0.6Cr, 0.15–0.25Mo
—	805–A17	0.15–0.2C, 0.4–0.6Cr, 0.15–0.25Mo, 0.4–.7Ni
—	805–A20	0.18–0.23C, 0.4–.65Cr, 0.15–0.25Mo, .35–.75Ni
BH13	BS4659	0.32–0.42C, 5Cr, 1.5Mo, 1.0V
BM2	BS4659	0.8–0.9C, 3.75–4.5Cr, 4.75–5.5Mo, 1.75–2.05V, 6–6.75W, 0.6 max Co.

2 Steel specifications American Iron & Steel Institute

AISI	Composition, main elements
6F2	.55C, 1.0Cr, 0.3Mo, 1.0Ni, 0.1V
304	0.08C max, 1.0Si, 2.0Mn, 18–20Cr, 8–10.5Ni
314	0.25C, 23–26Cr, 19–22Ni
316	0.8C max, 1Si, 2Mn max, 16–18Cr, 2–3Mo, 10–14Ni
410	0.15 C max, 12 Cr
440C	0.95–1.2C, 16–18Cr, 0.75Mo
4140	0.4C, 1Cr, 0.2Mo
4340	0.4C, .8Cr, .3Mo
4620	0.2C, 0.5Mn, 1.8Ni, 0.25Mo
6150	0.5C, 1Cr, 0.15V
8620	0.2C, 0.5Cr, 0.6Ni, 0.2%Mo
A2	1C, 5Cr, 1.2Mo, 0.4V
B1112	0.13C max, 0.9Mn, 0.2S free cutting
C1015	0.15C, 0.4Mn
C1020	0.2C, 0.4Mn
C1030	0.3C, 0.8Mn
C1040	0.4C, 0.8Mn
C1045	0.46C, 0.8Mn
C1060	0.6C, 0.8Mn
C1090	0.91C, 0.8Mn
D2	1.5C, 12Cr, 1Mo
D5	1.5C, 12.5Cr, 1Mo, 0.35Ni, 0.5V, 2Co
D6	1.5C, 12Cr, 1.0Mn
D7	2.3C, 12Cr, 1Mo, 4V
H13	0.35C, 5Cr, 1.5Mo, 1V
M2	0.85C, 4.0Cr, 6.3W, 5.0Mo, 1.9V
M7	1Cr, 4Cr, 8.7Mo, 1.7W, 2V
O2	0.95C, 1.6Mn, 2Cr*, 0.15V*, 0.3Mo*, *optional

3 Steel specifications American Society of Automotive Engineers

SAE	Composition, main elements
1137	0.36C, 1.5Mn
3135	0.35C, 0.6Cr, 1.2Ni
4140	0.4C, 1Cr, 0.2Mo
5115	0.15C, 0.8Mn, 0.8Cr
9310	0.1C, 1.2Cr, 3.2Ni, 0.1V
86830H	0.27–.33C, 0.35–0.65Cr, 0.15–0.25Mo, 0.35–0.75Ni, .0005 min B
G4000	Grey cast iron

4 Steel specifications German Standards

German	Composition, main elements
42CrMo4	0.38–45C, 1Cr, 0.6Ni, 0.2Mo
61CrSiVS	0.57–0.65C, 0.6-0.9Mn, 1–1.3Cr, 0.7-.12V
100Cr6	.95–105C, .25–0.4Mn, 1.4–1.7Cr
105Cr6	1.0–1.1C, 1.4–1.7Cr
105WCr6	1.06C, 1.0Cr, 1.2W
145Cr6	1.4–1.6C, 0.5–0.7Mn, 1.3–1.5Cr
C15	0.15C, SiMn
C45W3	0.45C, 0.7Mn
C/K15EH	0.15C, 0.45Mn
X22CrMoV12/1	0.2–.26C, .3Si, .6Mn, 12Cr, 1.2Mo, 1Ni, 0.35V
X40Cr13	0.4–5C, 12–14Cr, 1.0Ni
X45CrSi9/3	0.45C, 3.0Si, 9Cr
X155CrMoV12/1	1.55C, 0.4Mn, 12Cv, 0.7Mo, 1.0V

5 Other steel specifications

	Source	Composition, main elements
21–4–N	Jessop	0.55C, 9Mn, 21Cr, 4Ni, 0.4N
ASP23	Stora	1.2–1.7C, 4.2Cr, 5.0Mo, 6.4W, 3.1V
ASTM 369 FP1		0.15C, 0.7Cr, 0.5Mo
C45	Italian	0.45C, SiMn
C100	Italian	1C, 0.42Mn
HP9-4-20	ASTM A646	0.2C, 0.75Cr, 9Ni, 1Mo, 0.09V, 4.5Co steel
Nitralloy N	Latrobe	0.25C, 1.25Cr, 1.1Al, 2.5Mo, 5.5Ni
Vasco 7152	Vanadium alloys	1.7C, 17.4Cr–steel

6 Cobalt alloy specifications

	Source	Composition, main elements
Cobalt alloys		
Stellite 1	Cabot Corp.	2.5C, 3.3Cr, 1.3W, – Co alloy
Stellite 6	Cabot Corp.	1C, 26Cr, 5W, – Co alloy
Stellite 12	Cabot Corp.	1.8C, 29Cr, 9W, – Co alloy
Stellite 20	Cabot Corp.	2.5C, 3Cr, 18W, – Co alloy
Stellite 21	Cabot Corp.	0.25C, 27Cr, 2.5Ni, 4.5Ni, 4.5Mo, 2Fe, 0.007B Co alloy
Stellite 156	Cabot Corp.	28Cr, 1.6C, 1.1Si, 1Mn max, 1Mo max, 3Ni, 4W, Co alloy
Stellite 157	Cabot Corp.	22Cr, 0.1C, 1.6Si, 1Mo max, 2Fe max, 2Ni max, 2–4B, 4.5W, Co alloy
Ti-coat T93	Metallurgical Industries Inc	8C, 20Cr, 24Ti, 0.6Si–Fe
Ti-coat T930	(As above)	7C, 20Cr, 24Ti, 3W, 0.8Si, Co alloy
Tribaloy T100	Du Pont	35Mo, 10Si, Co-alloy
Tribaloy T400	Du Pont	8Cr, 28Mo, 2Si, Co alloy
Tribaloy T800	Du Pont	17Cr, 28Mo, 3Si, Co alloy

7 Nickel alloy specifications

	Source	Composition, main elements
Nickel alloys		
Deloro 40	Cabot Corp.	7.5Cr, 3.5Si, 1.2B, 1.0Fe-Ni alloy
Deloro 50	Cabot Corp.	10Cr, 4Si, 1.5B, 4.0Fe-Ni alloy
Deloro 60	Cabot Corp.	15Cr, 4.5Si, 3B, 4.5Fe-Ni
Hastelloy C	Cabot Corp	0.12C, 16Cr, 17Mo, 4W, 5Fe, 2.5Co Ni alloy
Haynes 711	Cabot Corp.	2.7C, 27Cr, 8Mo, 23Fe, 12Co, 3W,-Ni alloy
Haynes 716	Cabot Corp.	1.1C, 26Cr, 3Mo, 29Fe, 11Co, 3.5W, 0.5B, -Ni alloy
MCrA1Y	Alloy Metals Inc.	Ni-17Cr, 6Al, 0.5Y
Nimonic 75	International Nickel	20Cr, 0.4Ti, 5Fe-Ni alloy
Nimonic 80A	International Nickel	0.1C, 20Cr, 2Ti, 5Fe, 1Al, 2Co, -Ni
Nimonic 81	International Nickel	0.05C, 30Cr, 1.8Ti, 0.9Al, 1.0Co, -Ni
Tribaloy T700	Du Pont	15Cr, 32Mo, 3Si-Ni alloy

8 Miscellaneous alloy specifications

	Source	Composition, main elements
Other alloys		
A1MgSi	Alcan	0.7Mg, 0.4Si, A1 alloy
C2	US-'C'- Grade system	TiC 45% vol, 0.6C, 3Cr, 3Mo-steel matrix sintered carbides
Deloro 90	Cabot Corp	27Cr, 2.7C, 1.75Ni-Fe alloy
IMI 314	IMI	10.7Sn, 4.1Mo, 2.3Al, 0.2Si-Ti alloy
IMI 318	IMI	6Al, 4V-Ti alloy
IMI 680	IMI	4Al, 4Mn-Ti alloy
M10	ISO	Sintered carbide classification according to use—turning at medium or high cutting speeds
M15	ISO	Sintered carbide classification according to use—turning at medium cutting speeds
Ni-Hard	International Nickel	3C, 6Ni, 8Cr Martensitic cast iron

Sources of information on material standards

WOOLMAN, J and MOTTRAM R A
The Mechanical and Physical Properties of the British Standard En Steels (BS970–1955) Pergamon Press, 1967.

R B ROSS
Metallic Materials Specification Handbook—3rd Edition
E F Spon, London, 1980.

Iron & Steel Specifications
British Steel Corporation—Chorley & Pickersgill, 1974.

Woldman's Engineering Alloys
American Society for Metals, 1979.

Stahl Eisen Liste—2 Auflage—Verein Deutscher Eisenhuttenluete—Dusseldorf 1967.

APPENDIX 3

Hardness Measurements

Most of the hardness values are quoted as Vickers Numbers (HV). In the Vickers test a diamond pyramid indenter with a square base and having an angle of 136° between opposite faces is forced by a load F into the material under test. The size of the indentation is then measured. The Vickers hardness number is obtained by dividing the load F in kilogrammes force by the area of the indentation in square millimetres.

In macrohardness testing, the British Standard[1] specifies that the load should lie in the range 1–1000 kgf. In microhardness testing the ASTM standard[2] specifies that the load should lie in the range 1–1000 gf. As hardness values can vary with the applied load both standards recommend that the load used be stated after the hardness value. However, in many of the earlier reported values of macrohardness the load is not stated. In microhardness testing the size of the indentation can be of a size comparable with the individual phases in a multiphase material and values therefore vary with both load and test position.

In the Rockwell C test a conical diamond indenter with an angle of 120° is used and the depth of penetration under a standard load measured. The approximate Vickers equivalents (HV) of Rockwell C values (HRC) are given in the accompanying table.

In the Brinell test a hardened steel ball is used as the indenter. Brinell hardness values (HB) are the same as Vickers values up to 300. In harder materials Brinell values are lower than Vickers values because of deformation of the indenter.

References

1 British Standard 427 Part 1 1961.
   Method for Vickers Hardness Test.
   Part 1—Testing of Metals.

2 ASTM Standard E384–73 (Re-approved 1979).
   Standard test method for Microhardness of Materials.

*Approximate conversion of hardness values for steel**

Hardness		Tensile strength	
HV	HRC	MN/m^2	tons/in^2
940	68.0	—	—
920	67.5	—	—
900	67.0	—	—
880	66.4	—	—
860	65.9	—	—
840	65.3	—	—
820	64.7	—	—
800	64.0	—	—
780	63.3	—	—
760	62.5	—	—
740	61.8	—	—
720	61.0	—	—
700	60.1	—	—
690	59.7	—	—
680	59.2	2270	147
670	58.8	2239	145
660	58.3	2192	142
650	57.8	2162	140
640	57.3	2131	138
630	56.8	2100	136
620	56.3	2054	133
610	55.7	2023	131
600	55.2	1992	129
590	54.7	1961	127
580	54.1	1930	125
570	53.6	1884	122
560	53.0	1853	120
550	52.3	1822	118
540	51.7	1791	116
530	51.1	1744	113
520	50.5	1729	112
510	49.8	1683	109
500	49.1	1652	107

Hardness		Tensile strength	
HV	HRC	MN/m^2	tons/in^2
490	48.4	1606	104
480	47.7	1590	103
470	46.9	1544	100
460	46.1	1513	98.0
450	45.3	1475	95.5
440	44.5	1444	93.5
430	43.6	1405	91.0
420	42.7	1374	89.0
410	41.8	1343	87.0
400	40.8	1312	85.0
390	39.8	1274	82.5
380	38.8	1243	80.5
370	37.7	1204	78.0
360	36.6	1173	76.0
350	35.5	1143	74.0
340	34.4	1112	72.0
330	33.3	1073	69.5
320	32.2	1042	67.5
310	31.0	1004	65.0
300	29.8	973	63.0
290	28.5	934	60.5
280	27.1	903	58.5
270	25.6	880	57.0
260	24.0	849	55.0
250	22.2	818	53.0
240	20.3	787	51.0

*Not applicable to non-ferrous metals or to case-hardened steels

HV —Vickers
hardness

HRC—Rockwell C
hardness

Index

Abrasives	4, 6, 7
Abrasive wear (see also Wear)	2, 4
influence of hardness on	6
of laser hardened surfaces	5, 2
of sprayed coatings	134, 139, 140
of various coatings	81, 82, 107
test methods	A1
Activated reactive evaporation	110, 111, 112, 113
Adhesive wear (see also Wear)	2, 3
test methods	A1
AERE Harwell	115, 120
Aero gas turbines	31
coatings used in	32, 33
Agricultural machinery	146
Alumina coatings	
sprayed	128, 130
CVD	105
Alumina-titania, coatings	130
Aluminium alloys	
anodising	81, 82, 83, 92
electroplating	84
thermochemical treatments	78
Aluminium bronze, coatings	128
Amsler wear test	A1
results on nitrocarburised steels	68
Anodising	81, 92
Applications of surface treatments	22–44
Arc wire spraying	129, 130
Austenitic high chromium irons	145, 146
Austenitic manganese steel	
applications	143, 144
compositions	144
Basalt linings	44
Bending fatigue strength (see fatigue strength)	
Bias sputtering	110, 115
Blast furnace sinter	
abrasive properties of	44, 45
Bond coats	124, 128
Bond strength	14, 15
of sprayed coatings	124, 137, 138
Boost diffusion (carburising)	59
Boride coatings	
morphology of	75, 76
hardness of	75
Boriding (or Boronising)	11, 75, 157
cemented carbides	76
gas	76
pack	76
paste	76
salt bath	76
Brazing (for hardfacing)	158
Broaches	145
Brush plating	81, 91, 92
Bulkweld process	156
Carbonitrides	67, 68
Carbonitriding	10, 60
gas	61, 62
salt bath	60

Carburising	10, 55
atmosphere control	58
fluidised bed	59
gas	58
liquid	57, 58
low pressure	59
pack	56, 57
plasma	60, 61
processes	56
salt bath	57
steels	56, 57
using nitrogen based atmospheres	58
Case depth	
effect on load-carrying capacity	71
Cast irons	
high chromium	145, 146
Cavitation erosion	2, 8
Cemented carbide cutting tools	33
effect of coatings on performance	34, 35
Cemented carbide platelets	45, 46
Chemical treatments	12, 94–101
Chemical vapour deposition (CVD)	12, 22, 103–109
applications of coatings (see also cemented	
carbide cutting tools)	33, 107
reactions	104
properties of coatings	106, 107
Chromium boride	157
Chromium carbide coatings	128, 130
Chromium, electroplated	81, 83, 84
applications	86
from trivalent baths	86
on piston rings and cylinder liners	26, 27, 28
properties	83, 84, 85, 86
variation of hardness with temperature	85
wear resistance	81, 82, 85
Chromium oxide coatings	
sprayed	130
slurry	94, 101, 102
Chute linings	44, 146
wear of	44, 45
CLA—centre line average	
definition	4
Cobalt alloys	
electrodeposited	81, 87, 88, 91
with added particles	32, 90, 91
for hardfacing	147, 148
Coke	
abrasive properties of	44
Combustion gun processes	12, 124, 125, 126
Component shape,	
effect in process selection	14, 15
Component size	
effect in process selection	16
Composite coatings	
cobalt—chromium carbide electroplated	32, 90, 91
nickel—silicon carbide electroplated	90
nickel—ptfe electroless	97, 98
nickel—silicon carbide electroless	97, 98, 99
by welding	146, 147
Compressor blades	31, 128, 132
Conforma clad	158
Contact fatigue	7, 20, 71
test methods	A1
Copper alloys	
thermochemical treatment of	78

Copper—electroplated	87
Corrosion resistance of coating materials	13, 17, 20
Corrosive wear	2, 9
Costs	
of solid tiles	19
of sprayed coatings	136
of surface treatments	19
of thermochemical treatments	72, 73
of welding processes	158, 159
Cutting tools (see cemented carbide cutting tools or high speed steel cutting tools)	
Cylinder liners	26
Delsun process	78
Density of sprayed coatings (see porosity)	
Deposition rates	16
Detonation gun (D-gun) process	12, 13, 14, 126, 127
coatings	128
Dowty Electronics plc	115
Drills	
effect of coatings on performance	36, 37, 38, 39
Drop forging dies	148
Dry rubbing wear, test method	A1
Earth moving equipment	42, 43, 44
hardfacing of	144, 146
Electrochemical treatments	12, 13, 81–93
equipment and process	82, 83, 84
Electrodeposited coatings, properties and applications	81, 82, 83
Electroless plating (see also Nickel-electroless)	12, 94, 95
applications	99
with added particles	95, 97, 98
Electron beam	
evaporation	111, 112
hardening	53, 54
Electroslag welding	157
Elowig process	50, 51
Endothermic gas	58
Environmental factors, influence on process selection	20
Erosion	
cavitation	2, 8
particle impact	2, 5
Erosive wear rate, effect of impact angle on	7
Evaporation processes	111, 112, 113
Exhaust valves	30, 31
Falex (Faville-Levally) test	A1
test results, electroless nickel	97
test results, electroplated chromium	97
Fan blades	31
Fatigue strength, bending	
effect of carburising	69, 72, 73
effect of electroless plating	97
effect of electroplating	18
effect of nitrocarburising	68, 69
effect of plasma nitriding	67
effect of sprayed coatings	138
effect of various treatments	17, 18
of gears	23
of through hardened steels	73
Fatigue strength, contact	A1, 71
Fatigue strength, hammer tests	A1
Fatigue wear	2, 7
Flame hardening	50
Flame spraying (see thermo-spraying or plasma spraying)	
Flux coatings	153
Flux-cored welding wires	154
Fretting	2, 8, 31
test methods	A1
Fused salt baths	
for carbonitriding	60
for carburising	57
for electroplating	92
for nitrocarburising	67
low toxicity	68
Fuse-weld process	125
Fused sprayed coatings	124, 125
FZG gear machine	A1
Gears	22
design data	23
scuffing, effect of various treatments	23, 24, 25
scuffing, test method	A1
thermal and thermochemical treatments of	23, 24, 25
Glow discharge processes (see plasma carburising or plasma nitriding or ion plating)	
Gold alloys, electroplated	88, 89
Gouging abrasion	2, 6, 7
Grinding abrasion	6
test method	A1
Grit blasting	123
Hammer wear	31
methods	A1
results	33
Hard anodising (see anodising)	
Hard chromium plating (see chromium plating)	
Hardening (see thermal hardening)	
Hardfacing	142–160
exhaust valves and seats	28, 29, 30
rock and earth engaging tools	42, 43, 44 146
comparison of methods	150
comparison of properties	144, 145
cost of	158, 159
definition	142
equipment	150–158
materials	143–150
process selection	142, 150
Hardness	
factor in treatment selection	14
influence on abrasive wear rate	6
influence on low stress abrasion resistance	4
influence on porosity on	15
influence on rolling contact fatigue strength	71
methods of measurement	A2
of steels with varying carbon content	56
of various materials	13, 75, 106
Health and safety considerations	
during nitrocarburising	68
during spraying	136, 137
during welding	151
High frequency resistance heating	49
High speed steel cutting tools	36, 37, 38, 39
Hot working tools	145
Hydrogen embrittlement	84
IAE gear machine	A1
Impact resistance	14, 145, 146
Induction hardening	48
effect of frequency	49
suitable steels	48
Inert gas spraying	134
Infra-red gas analysis	
for carburising atmospheres	58
Ion beam mixing	120
Ion beam processes	120
Ion bombardment	110, 113, 116
Ion implantation	110, 120, 121
Ion nitriding (see plasma nitriding)	
Ion plating	110, 116, 117
using multi-arc sources	118, 119

169

Iron, electroplated	88
Iron nitride	
ε and γ′	63, 65, 67
Jet-Kote process	131, 132
Kaman Sciences Inc.	101
Kawasaki Steel	157
Laser hardening	25, 26, 51, 52
Laser hardfacing	160
Lead alloys, electroplated	89, 90
Lead coatings	
ion plating of	119
LFW-1 test method	A1
Liquid carbonitriding (see fused salt bath)	
Liquid carburising (see fused salt bath)	
Liquid nitrocarburising (see fused salt bath)	
Machining	
effect of coatings on tool performances	34, 38
Machining wear	2, 4
test methods	A1
Maglay process	157, 158
Magnetrons (in sputtering)	110, 114
Manganese steels (austenitic)	143
applications	144
compositions	144
Manual arc welding	153
Martensitic high chromium irons	145, 146
Martensitic steels	144, 145
Metal cutting tools (see cermented carbide and high speed steel cutting tools)	
Metal forming tools	145
effect of coatings on performance	40, 41, 107
Metal inert gas welding	155, 156
Metalliding	11
Metallising (see wire spraying)	
Mild wear	2
Mining equipment	144, 146
Molybdenum coatings	
for piston rings	129
Multi-Arc (Europe) Ltd.	118
Nickel alloys	125, 127, 147, 148, 149
Nickel aluminide coatings	124, 129
Nickel boride	97
Nickel-boron alloys	95, 96, 97, 148
Nickel-electroless	94, 95, 96
applications	99
properties of	95, 96, 98
with added particles	97
heat treatment of	95
Niflor	98
Nitemper process	68, 69
Nitralloy steels (see nitriding steels)	
Nitrides (see iron nitride and titanium nitride)	
Nitriding	10, 61–65
effect of tempering on hardness after	64
gas	62
of gears	24
plasma	63, 66
potential	62, 65
steels	64
Nitrocarburising	10, 65–70
gas	68
liquid	67
low pressure	65
of gears	24
process developments	69
salt bath	67
steels	68

Nitrotec process	70, 71
Noskuff process	60
Open arc welding	154
Operational factors,	
influence on coating selection	20
Ore handling equipment	146
Oxide coatings, plasma sprayed	130
Oxidation resistance	15
Oxy-acetylene welding	151, 152
Particle impact erosion	2, 5
Paste fusion (welding) process	157
Percussive wear	2, 8
Phosphating	12, 99, 100, 101
applications	99, 100
properties	101
Physical vapour deposition (PVD)	12, 110–122
comparison of processes for	110
Pin and disc test methods	A1
Piston rings	26–29
Pitting	7, 8
Plasma arc weld surfacing	150, 152, 153
Plasma carburising	60
Plasma sprayed coatings	130
carbide-metal cermets	130
metals	129
oxides	130
wear resistance of	134, 138, 139, 140
Plasma spraying	12, 130, 131
at low pressures	132, 133
with inert gas shroud	134, 135
Poppet valves	28, 29, 30
Porosity, of coatings	15, 124
Powder spraying (see thermo-spraying)	
Powder welding	
oxy-acetylene	152
plasma arc	152
Pulverised fuel,	
handling equipment	45, 46
Pulverising hammers	145
Rates of deposition and layer formation	16
Reactive evaporation	110, 111
Reactive sputtering	114
Reamers	145
Resistance heating	49
Residual stresses	
in heat treated steels	72
Rock and earth engaging equipment	42, 43, 44
Rokide process	129, 138
Scuffing	2, 3, 4
test methods	A1
Severe wear	2, 3
Shear blades	145
Shielded metal arc process	153
Silver alloys, electroplated	89
Sinter	
abrasive properties of	44, 45
Sliding wear	2, 4
Solid tiles	12, 19, 45, 46
Specifications,	
weld deposits	143
materials	A2
Sprayed coatings	123–141
bond strength	124, 137, 138
costs	136
effect on fatigue strength	138
finishing, surface	134
limitations	123
porosity	124
quality control	136

thickness	124	PVD coatings	116
wear resistance	138, 139, 140	Toyota Diffusion (or TD) process	11, 40, 41, 76
Spraying processes (see also D-gun process,		comparison with nitrocarburising	77
plasma spraying, thermo-spraying)	12, 123–135	rate of formation of carbide	77
Spreader techniques, weld deposits	42, 43	Tribaloy alloys	129, 148
Sputtering process	110, 113–116	Tribomet coatings	90
using magnetrons	114	Tube Investments plc	115
Steels (specifications)	A2	Tubular welding wires	154
Steering gear (laser hardening of)	25, 26	Tufftride	24, 78
Stellite	A2, 129, 147,	Tungsten	
	148, 149	CVD coatings	105, 106, 108
Strip electrodes	155, 158	Tungsten carbide	
Submerged arc welding (see welding)		CVD coatings	105
Substrates		composites for welding	146
temperatures during processing	14, 15	Tungsten carbide-cobalt	
considerations in welding	151	boriding of	76
Sulf BT	11, 24	sprayed coatings	128, 130
Sulfinuz process	11, 68	wear tiles	45, 46
Sulfocyaniding	68		
Sulfurising	68		
Sulphidising	68	Ultimate resilience characteristics	8
Surface coatings	12	Union Carbide Corporation	127
Surface heat treatments	10, 11		
Surface preparation			
electroless plating	94	Valves	28, 29, 30, 148,
electroplating	84		149
spraying	123, 124	Vanadium carbide	
Surface treatment processes	10	composites (in welding)	147
comparison of	10–20	Vapour pressure of metals	111
effect on substrate	17,18		
factors affecting selection	13–20		
Sursulf process	11, 68	Wear	
Synthetic fibre processing machines	32, 33	abrasive	2, 4, 5
		adhesive	2, 3
		by coke and blast furnace sinter	44, 45
Taber abrasion test method	A1	cavitation	2, 8
test results, electroless nickel	96	corrosive	2, 9
wear index	A1	delamination	2, 8
Temperature of substrate	14, 15	equations	9
TD (see Toyota Diffusion)		erosive	2, 5, 7
Tenifer process	68	fatigue	2, 7
Test methods, for wear,	A1	fretting	2, 8
Thermal evaporation	110, 111, 112	gouging abrasive	2, 6
applications of coatings produced by	113	influence of mineral hardness	4, 5, 6
Thermal hardening	10, 48–54	influence of impact angle	5, 7
by electron beam	53, 54	influence of surface hardness	5
by flames	50	machining	2
by high frequency resistance heating	49	mild	2
by induction	48	of CVD coatings	107
by lasers	51, 52, 53	of electroplated Au-Co alloys	89
by TIG melting	50, 51	of electroplated chromium	81, 82, 85
suitable steels	48	of sprayed coatings	134, 139, 140
Thermochemical treatments	55–80	percussive	8
comparison of	70–73	processes	2–9
cost of	73	severe	2, 3
factors affecting choice of	74	sliding	2, 4
Thermospraying	124–127	test methods	A1
Thermosprayed coatings	125	three body abrasive	2, 6
Thickness of case, thermochemical (see case		tiles, tungsten carbide	45, 46
depth)		Wear resistant materials	13
Thickness of coatings	16	Welding processes	12, 150–157
Three body abrasion	2, 6	electroslag	157
Titanium alloys	78	manual metal arc	153
boriding of	79	metal inert gas (MIG)	155
electroplating of	84	open arc	154
nitriding of	78	oxyacetylene	151, 152
oxidation of	78	plasma transferred arc (PTA)	152, 153
salt bath treatments of	79	submerged arc	155, 156
Titanium carbide		tungsten inert gas (TIG)	152
CVD coatings	104, 106	White layer (in nitriding)	61, 62, 63, 65
PVD coatings	113	Wire drawing dies, boriding of	76
composites (in welding)	147	Wire spraying	128, 129
Titanium carbonitride		Wire sprayed coatings	128
CVD coatings	105		
Titanium nitride			
CVD coatings	104	Zinal process	78

Printed in the UK for HMSO
(2249/85) Dd738815 C15 6/86 HP Ltd So'ton G381